META Morphing

VIVIAN SOBCHACK
EDITOR

Visual Transformation
META Morphing and the
Culture of Quick-Change

University of Minnesota Press
Minneapolis • London

Published by the University of Minnesota Press
111 Third Avenue South, Suite 290
Minneapolis, MN 55401-2520
http://www.upress.umn.edu

Library of Congress Cataloging-in-Publication Data

Meta-morphing : visual transformation and the culture of quick-change / Vivian Sobchack, editor.
p. cm.
Includes bibliographical references.
ISBN 0-8166-3318-5 (hc.) — ISBN 0-8166-3319-3 (pb.)
1. Morphing (Computer animation) I. Sobchack, Vivian Carol.
TR897.75.M48 2000
700′.415—dc21 99-044863

Printed in the United States of America on acid-free paper

11 10 09 08 07 06 05 10 9 8 7 6 5 4 3 2

For Christopher, my son—

the most thoughtful quick-change artist I know

Contents

Acknowledgments

Many people and institutions contributed both formally and informally to the realization of this volume. I must thank the Academic Senate of the University of California, Los Angeles, for funding a grant that provided research assistance. Because of their support, I was able to take advantage of the indefatigable legwork and invaluable counsel of Kevin Fisher, who not only contributed an essay to the volume but also assisted me from beginning to end. I cannot thank him enough. Louise Krasniewicz was unstintingly helpful and extremely generous of both time and spirit in constructing "morph" illustrations and allowing me to use her photographs for this book. Judy Conner of Pacific Data Images was also most helpful in this regard. Thanks also go to those anonymous reviewers who read the manuscript for the University of Minnesota Press and offered invaluable suggestions and to all those better-known colleagues who stimulated my thoughts about morphing and electronic culture in many memorable conversations—chief among them Peter Lunenfeld, Lev Manovich, and Stephen Mamber. Finally, I want to express my gratitude to Micah Kleit (formerly of the University of Minnesota Press), whose excitement for this project fueled my own. My thanks also to William Murphy, Jennifer Moore, and Robin A. Moir of the University of Minnesota Press, who have always been helpful, gracious, and—most of all—patient.

Introduction

VIVIAN SOBCHACK

My mind is intent on singing of shapes changed into new bodies.

Ovid, *Metamorphoses*

They imagined it as . . . a misshapen monster which devoured all things . . . the prime matter . . . which, formless itself, is greedy . . . and devours all forms.

Vico, *The New Science*

Computergraphic "morphing" fascinates us—even in the most banal of its visible figurations. Amid myriad forms and articulations of computer-graphic representation, there is certainly something particular about the "morph" that compels our contemporary attention. Against the ground of (and sometimes grounding) the photo-realisms of film and television, its effortless shape-shifting, its confusions of the animate and inanimate, its curiously static movement, its queerly hermetic liquidity, its homogenizing consumption of others and otherness, are uncanny—uncanny not only in the sense of being strange and unfamiliar but also in the sense of being strangely familiar. Indeed, one could argue that at this particular historical moment and in our particularly digitally driven American context, the morph fascinates us not only because of its physical impossibility and strangeness but also because its process and figuration seem less an illusionist practice than both a presentational mode and an allegory of late capitalist "realism."

Like that which is uncanny, morphing generates our physical and cultural "double"—some radically other "familiar" whose visible image (particularly when it's a metal-morph like the T-1000) not only "reflects" us but also "renders" and "clarifies" us.[1] As our physical double, the morph interrogates the dominant philosophies and fantasies that fix our embodied human being and constitute our identities as discrete and thus reminds us of our true instability: our physical flux, our lack of self-coincidence, our subatomic as well as subcutaneous existence that is always in motion and ever changing. As our cultural double, the morph enacts our own greedy and effortless consumerism—at the same time as it terrifies us with reflected images of our own consumption. Making formally visible the very formlessness at its center, the morph also makes visible our national and political sense that although there is power, there is no center, that centers no longer have substance (at least as we once believed), that centers do not hold. Exuberant and liberating in its democratic lack of hierarchical attachment to any privileged form of being while bent on some ultimate totalitarian mastery of all forms, exuding its own freedom while globally incorporating and homogenizing that of all in its path, the morph seems to double the dramatic actions of both our nation and our technoculture.[2] At the same moment, its very fluidity destabilizes dominant Western metaphysics (primarily focused on essences, categories, and identities, including those of gender and race) and dramatizes instead a "process metaphysics" that is less about "being" than about "becoming." (Introducing a similar critique of Western metaphysics, Michel Foucault writes: "Do not ask who I am and do not ask me to remain the same.")[3] Morphing's dramatic emphasis on process thus foregrounds not only metaphysical but also political contradictions. That is, it threatens to dissolve dominant fixations of "American" identity while also appealing to their very mythos and grounding in the American ideal of social mobility and the "be all that you can be" mutability of the "self-made man."

Morphing is also uncanny (as are many other historical renderings of metamorphosis) not only because of its "supernatural" acceleration of the processes of change and transformation but also because of their (usually) graceful and effortless accomplishment—the strangeness of this speed and grace relative, of course, to our own temporal rhythms and to the resistance of worldly others and things we encounter daily in our mortal Western lives. Nonetheless, in the historical present of a particularly electronic "American" culture and economy, the effortless "quick-change" we may, as physical beings, regard as the uncanny "otherness" of the morph is also strangely familiar and culturally self-reflexive—this because the morph, in the microcosmic fluidity of its figural transformations, enacts the more macrocosmic and fluid quick-changes not only of the computer and entertainment industries that generated it but also of the larger metamorphic technosphere in which we all live. Which is to say that the cultural ubiquity of "quick-change" has itself increased and accelerated from the mid-nineteenth century on—thanks to changes in industrial technology and modes of production, new communications technologies such as the telegraph and telephone, and also the transformational "magic" of photography, cinema, and now the computer. Thus, for all its uncanniness, "quick-change" has been a primary and familiar fact of daily American life since the turn of the century (and, in the cities, before it). As Cecelia Tichi notes of the period: "Speed and the belief in cultural acceleration were proclaimed from every quarter to be, for better or worse, *the* defining characteristic of the United States."[4] This defining characteristic has not changed, but our sense of "quick-change" itself has quickened.

It is no wonder, then, that morphing and the morph fascinate us. Morphing is not merely a novel computergraphic mode of figuration, nor is the morph merely a novel narratological figure. Both are novel—and specifically historical—concretions of contemporary confusions, fears, and desires and both, whether visible or invisible in their use, allegorize the quick-changes, fluid movements, and inhuman accelerations endemic to our daily lives. Thus, as a form (both of figuration and figure) that is "carried beyond" itself and "across" different realms of our present existence and culture, the morph is not only *meta-morphic* in its shape-shifting formlessness that greedily "devours all forms"; it is also *meta-phoric* in its inherent tropological movement and its historically substitutive activity.[5] Hence the specific title and project of *Meta-Morphing: Visual Transformation and the Culture of Quick-Change.*

Given this introductory gloss, it should be clear why morphing is of sufficient interest to warrant a collection of essays devoted to both its genealogy and the historical, cultural, and aesthetic significance of its contemporary practice. This particular volume, however, emerged from both an event and a conviction. The event was a panel I organized and chaired, also called "Meta-Morphing," for the 1996 Society for Cinema Studies annual conference, held that year in Dallas, Texas. Enthusiastic reception of the topic and the lively discussion that followed convinced me that I was only one among many media scholars fascinated by this novel form of digital figuration.[6] The conviction (one also informing the panel) was—

and still is—that for those of us interested in our current "electronic culture" and its "new technologies," it is now time to move beyond general analyses and broad critique of our digital culture to a more focused theory and criticism that distinguishes among a variety of quite specific digital articulations. Certainly, large and homogenizing claims about the post-industrial West's novel historical and cultural "electronic" moment have been absolutely necessary to an initial mapping of a new field of inquiry, but their heuristic power has steadily diminished: where once they excited us, they now seem somewhat repetitive and far too broad in focus. What is historically and technologically novel about digitization is precisely its unique capacity to translate all other media representation into a homogeneous algorithmic mode of expression; nonetheless, we have come to recognize that digital representations are extraordinarily heterogeneous in form, diverse in function, and specific in practice. Thus as Stephen Prince has pointed out in relation to digital representation (particularly as it interacts with and changes photographically based, photo-realist cinema):

> Because the digital manipulation of images is so novel and the creative possibilities it offers are so unprecedented, its effects on cinematic representation and the viewer's response are poorly understood. . . . Film theory will need to catch up to this rapidly evolving new category of imaging capabilities and grasp it in all of its complexity.[7]

The goal of *Meta-Morphing,* then, is to grasp the historically novel articulation of morphing in all its digital complexity. The collection thus looks both at the material and technical aspects of a new technologically constituted form of figuration and figure and at the metaphysical and metaphoric aspects of this new form as it substitutes for an older imagination of transformation and "quick-change" in time and space and substance. Here, morphing is grasped both as a *digital practice* that is historically and materially discrete and specific and as a *cultural imaginary* that links a present digital practice to a much broader history and tropology of metamorphosis and its meanings. On the one hand, the essays to follow take up morphing as a historically novel representational practice whose specific material means radically interrogate certain traditional notions of coherence and self-identity in space and time, of narrative and character, of evolution and devolution. (In my own essay, for example, I discuss how the morph takes the *bildung* out of the *bildungsroman.*) On the other hand, however, the essays take up morphing as a particular representation of metamorphosis that expresses a cultural "desire" susceptible to an archaeology and genealogy. That is, digital morphing both turns us backward to classical and cross-cultural mythology and magic, to explorations of change over time as conceived and figured by premodern scientists, mathematicians, and artists, to the quick transformations performed in theatrical magic shows at the turn of the century, and to the early cinema of "attractions" and animated and "trick" films, and also turns us outward into the present cultural moment to consider its connection to much contemporary "body art" and to our current belief in the "artifice" of a body that can be transformed not only through digitization but also through "naturalized" cosmetic surgery. Collectively, then, the essays in this volume take a critical look at digital morphing both in and beyond its immediate electronic context in an attempt to explore its continuities and discontinuities with earlier forms and figures of "marvelous" transformation.

Obviously, one could attempt to trace various conceptions of metamorphosis as marvelous "quick-change" across the breadth of history and through a variety of cultures, but that would be an impossible task for any one book. Thus, although there are cross-cultural comparisons and broad historical references made in many of the essays, primary emphasis is on the twentieth century and on Western culture, most specifically American culture and media texts. As I have suggested earlier, such an emphasis may be justified by the accelerated technologization and digitization of culture in the United States as compared to other countries, and by a long-held popular belief in technology (now digital) that has always dazzled us with utopian visions wrought less, I would argue, by the logic (or rhetoric) of efficiency than by an aesthetics of power, speed, effortlessness, and grace.[8]

The essays in this volume look not only at film, video, television, theater, and performance art but also to the history of mythology, mathematics, the visual arts, technology, literature, and philosophy, and they range across a wide variety of disciplines and modes of inquiry and description. Nonetheless, they are united in an attempt to understand both what a general imagery of metamorphosis means historically to our culture and how that imagery and its meanings have themselves been transformed by the historically novel and specific computergraphic articulations of digital morphing. Thus many texts and themes are common and emerge as central to an exploration of the morph. It should be no surprise that most prominent among the texts discussed are Michael Jackson's music video *Black or White* and James Cameron's *Terminator 2: Judgment Day* (both 1991). Certainly, among prominent and recurrent themes are the *representation* and *performance* of rapid transformation in time and space and the "transmogrification" of substance. So, too, are the morph's associations with the practice and attractions of *magic*—whether such magic is associated with human dexterity, the mythic power of ancient gods, or the wizardry of digital technologies. Emphasizing the temporal qualities of "quick-change," a number of the essays also explore the morph's implications for *narrative*—not only in relation to plot structure and characterization but also in relation to the structure of spectatorial *identification* and the source of *narrative pleasure.* Many of the essays—most of them skeptical—also focus on the way in which morphing seemingly breaks down and collapses *boundary distinctions,* not only between ontological categories such as animate and inanimate or human and animal but also between epistemological categories and their historical and cultural fixations of *gender, race,* and *ethnicity.* Indeed, while several essays consider the morph's interrogation of any fixed notion of *"self-identity,"* others are concerned with the morph's supposed dissolution of gender, race, ethnicity, and *difference.* Some of these same essays are also critical of the morph for precisely the quickness of its change and thus its smoothing over or complete elision of *history.* Finally, to varying degrees, nearly all the essays discuss *metamorphic technologies:* most often the cinematic and the digital.

Meta-Morphing has thus been organized into three parts that loosely group the essays in relation to a specific set of entailed themes. The first part, "Metamorphosis, Magic, and Mythology," focuses on various ways in which material transformations in time and substance have been articulated to evoke wonderment of a nontechnological kind. Here metamorphosis is connected with a historical imagination of, and amazement at, the protean; with a sense of the ineffable, mysterious, and uncanny aspects of transformation; and with cosmologies more spiritual and less compartmentalized than our own.

Matthew Solomon's "'Twenty-Five Heads under One Hat': Quick-Change in the 1890s" begins part I. His essay looks at the protean and popular theatrical arts of "quick-change," "chapeaugraphy," and "shadowgraphy" before and at the turn of the century. These acts of metamorphic performance, he argues, were not simply precursors to the "attractions" and "magic" of the "quick-change" allowed by the new technology of cinema, which all too rapidly shifted "transformations from the register of performance to technology." Rather, these performed transformations demonstrated an embodied fluidity, a "plasmaticness," available to cinema, but not taken up by it except in certain cartoons (or, as Scott Bukatman later notes, in the films of performers such as Jerry Lewis). It is thus historically interesting that the cinema's recent incorporation of the digital has led the medium back to some of its underdeveloped possibilities and to the foregrounding of metamorphic performance.

Norman M. Klein's "Animation and Animorphs: A Brief Disappearing Act" picks up on this imbrication of performance, transformation, and "plasmaticness." In early animated films, as well as in the Fleischer cartoons of the 1930s and early 1940s, he finds not only self-reflexive demonstrations of the animators' "lightning-hand" dexterity but also articulated "traces" of transformation and movement that point to the gaps, fissures, spaces, and ruins that constitute the "blur" left by all this shapeshifting. Klein privileges these self-reflexive and "troubling" signs of metamorphosis (or "ani-morphs," as he calls them) because they represent the loss in-forming human temporality and transformation. Linking animorphic representation to the broader cultural context, Klein notes its disappearance from the disciplined and contained metamorphoses made popular by Disney animation but sees its return in more contemporary works such as the Quay brothers' *Street of Crocodiles.* Finally, he wonders to what extent the computational consciousness underlying computergraphic morphing can accommodate and represent the "lightning-hand" or ani-morphic reflexivity.

Louise Krasniewicz's "Magical Transformations: Morphing and Metamorphosis in Two Cultures" also reflects on computergraphic transformations of morphing and considers their current superficiality in contrast to the rich imagination of metamorphosis demonstrated by Mesoamerican culture. Her essay focuses on the religious and political dimensions of magical transformation for the Olmec shaman, who entered the spirit world transformed into a jaguar. Our own culture's parallel metamorphosis—the digital morph of a car, gassed up with Exxon, into a tiger—foregrounds not only the fact that our spiritual imagination has been supplanted by a technological imagination but also the latter's (unnecessary) impoverishment. By treating metamorphosis and morphing in a cross-cultural and comparative context, Krasniewicz draws attention to the aesthetic, ethical, and mythological potential of digital morphing insofar as it is, like the Olmec shaman, able to "to move laterally across categories" and undo hierarchical order and binary thought.

Marsha Kinder's "From Mutation to Morphing: Cultural Transformations from Greek Myth to Children's Media Culture" continues Krasniewicz's interests in the rich mythology of metamorphosis and in the "lateral movement across categories" that metamorphosis allows. This movement, however, is less spiritually than economically driven in contemporary culture: Kinder considers children's movies, television series, morphing toys, and "ancillary products," which not only figure but also construct new mythologies—while maintaining old gender and racial norms. The Teenage Mutant Ninja Turtles and the Mighty Morphin' Power Rangers serve as emblems of the broader morphology in which media forms "transform" and expand their own productions across entertainment genres and marketing categories as mythology. Like Krasniewicz, Kinder considers current morphology and mythology and its similarities and differences to the tropes of shape-shifting found in the literature of metamorphosis from Ovid to William Burroughs.

Part II is called "Transformation, Technology, and Narrative." Whereas the essays in part I give major emphasis to the magical and mythological imagination of protean transformation, the essays here are more focused on the specificity of morphing as both a digital practice and a novel form of figuration that emerges primarily from the scientific and technological imagination. The essays here are also most concerned with the way in which a novel form of visual representation allows a new way of seeing, adds a new dimension to the cultural Imaginary, and ultimately "transforms" philosophical thought and narratological structures.

Although digital morphing is mentioned in the previous essays, Mark J. P. Wolf's "A Brief History of Morphing" provides us with a necessary introduction to the specificity of digital morphing, warping, rendering, and compositing. In both historical and concrete terms, Wolf's unadorned and careful description of what digital morphing technically entails serves as grounding for more philosophical, cultural, and aesthetic reflections on the structure and meaning of this new form of figuration. Although many of this volume's readers may be familiar with digital image manipulation in general and morphing software in particular, many others who are fascinated by morphing as figuration and figure may know very little about how such image manipulation is accomplished and what is entailed in its practice that might have significant structural consequences for aesthetics and for narrative. Wolf's essay serves as concrete ballast to the more interpretive work that follows.

Whereas Wolf concentrates on digital morphing in relation to two- and three-dimensional images, Kevin Fisher's "Tracing the Tesseract: A Conceptual Prehistory of the Morph" takes us into the imagination and visual representation of the "fourth dimension"—transformation in space-time not its equivalent, but for us (in the third dimension) its visible sign. His essay focuses on historical variations of the visualization of the "tesseract," a term used generally to describe a four-dimensional extension of a three-dimensional object and here specifically to describe the essential structure of the digital morph. Moving across a range of conceptualizations of the "fourth dimension" that emerge in geometry, mathematics, physics, photography, painting, and cinema, Fisher traces the "prehistory" of the digital morph to its present moment and illustrates both its structure and its poetic force in two films that use morphing and the figuration of the fourth dimension in quite different but related ways: *Terminator 2* and *Heavenly Creatures.*

My own essay, "'At the Still Point of the Turning World': Meta-Morphing and Meta-Stasis," also looks at the structure of the digital morph, but with an emphasis on its philosophical significance and its consequent implications for conceptions of character and narrative. Initially, morphing is considered a novel form of figuration that accomplishes transformation and change in time in a structure that is quite different in temporality from such cinematic figurations as the cut, dissolve, or camera movement combined with long takes, all of which constitute temporal significance and eventful meaning in sequential terms. Morphing, I argue, is temporally reversible, palindromic, and its novel temporal structure and lack of hierarchical meaning has implications for our notions of human transformation and character and thus narratives of "becoming." Using Mikhail Bakhtin's historical discussion of novelistic chronotopes of "human becoming," the essay goes on to explore the use of the digital morph as a specific figure with specific temporal value in a selection of contemporary texts.

Part II ends with Roger Warren Beebe's "After Arnold: Narratives of the Posthuman Cinema," which continues the exploration of the morph as a novel narrative figure with significant structural effects. Grounding his discussion in *The Terminator* and *Terminator 2,* Beebe argues that digital morphing and the morph have transformed narrative and, particularly, the action genre and its primary figure, the action hero (typified by Arnold Schwarzenegger). Indeed, morphing and the morph have moved the action film not only beyond its previous emphasis on character (however stereotypical) but also beyond much emphasis on the "human"—or on stars. Thus morphing and the morph have also transformed the structure of spectatorial pleasure, serving much like Barthes's "punctum" to disrupt the spectator's traditional modes of identification with central human characters and to displace them onto posthuman dramatizations of technological *jouissance.* Beebe argues that—whether they use digital morphing or not—the morph and computergraphic technology have altered the shape of contemporary generic film narratives, enabling a "whole new form" in which, given current trends, Arnold may well not "be back."

Part III is entitled "Morphing, Identity, and Spectatorship," and it moves us to further consideration of digital morphing's interrogation of such notions as identity, identification, and spectatorship. All the essays, in one fashion or another, deal with issues of social performance and of the internal contradictions that perhaps no longer "split" subjects so much as shape-shift them into self-conscious instability. Given their emphasis on identity and identification, on "self" and "other" and their indistinction, three of the essays deal significantly with digital morphing and such defining discriminations as gender and race. In a fitting conclusion, however, the final essay moves to a consideration of boundary distinctions that no longer hold at a broader level of identification and performance: namely, those that have separated the "spectator" and the "text," and those that have separated media forms from each other.

Joseba Gabilondo's "Morphing Saint Sebastian: Masochism and Masculinity in *Forrest Gump*" begins part III by looking at the way morphing has been used in contemporary cinema simultaneously to reimagine American history and American masculinity so as to negate the violence historically attached to white male patriarchy and its will to power and domination. Gabilondo considers morphing not only in its spectacular but also in its "invisible" modalities and connects its effects in both modes to classical cinema's strategies of "suturing" the spectator into the text and into a structure of identification with its point of view and ideological position. Focusing on *Forrest Gump* precisely because of its transparent or "low" use of morphing and thus its "high" degree of suture, Gabilondo shows how the film uses morphing to overtly reproduce the relation of masculinity to history not only as kinder and gentler but also as masochistic—at the same time, using morphing not only to covertly erase past acts of sadism but also to enact present ones that quietly erase the threats posed to masculinity by race and gender.

The next essay takes another approach to the relationship between digital morphing and gender. Victoria Duckett's "Beyond the Body: Orlan and the Material Morph" reads the French performance artist's staged surgical operations as radical interrogations not only of cosmetic surgery and standards of female beauty but also of the technological imagination that would morphologically erase visible signs of the real and bodily costs of such metamorphoses. Using materials as diverse as medical disclaimers about computergraphic imagery and tabloid press pieces that reveal the morphological "secrets of the stars," Duckett considers the complexity of Orlan's project within the context of a digital culture that believes the rhetoric of its own metamorphosed images even as it legally disowns them or publicly reveals them.

Scott Bukatman's "Taking Shape: Morphing and the Performance of Self" also deals with performance of the self and the instability of self-identity. His wide-ranging essay moves from a consideration of the mutability and insecurity of the self as it has been articulated by nondigital artists as diverse as Marcel Proust and Jerry Lewis—and then moves to an extended analysis of digital morphology as it has been articulated by Michael Jackson and by Jim Carrey in *The Mask* in a specifically racial discourse. Maintaining that the morph is a "way of seeing" that exaggerates quotidian perceptions of continuity and discontinuity, Bukatman views the morph as a "slippery masquerade" of self that, at every turn, turns us into an "other" but also elides and represses history in the guise of "regenerative self-creation."

The final essay in the volume, Angela Ndalianis's "Special Effects, Morphing Magic, and the 1990s Cinema of Attractions," moves us out from the morphology and performance of self-identity to the broader metamorphoses that morphing in particular and digital technologies in general have meant for both traditional cinematic spectatorship and media forms. Her major focus is on *Terminator 2: 3D Battle across Time*, a seamless "multimedia" theme park attraction derivative—but also transformative—of its cinematic predecessor, *Terminator 2*. Ndalianis's analysis of this hyperbolic and technically "amazing" display of "hyperreal constructions" turns full circle back to considerations of "magic shows" and the early "cinema of attractions" of this volume's inaugural essay. It also demonstrates how current digital technologies have effected an increasing convergence of theater, film, and computergraphics so as to create an immersive and illusory environment in which "magic" is reversible with "method," and in which audiences "act" as much as they "spectate."

The major contribution of the essays in *Meta-Morphing* will be found in their refusal to reduce the digital morph to a mere technical effect of a technological practice or merely to contain its meanings within the aesthetic or semiological confines of the close analysis of a few representative texts. Meant for an audience who is interested in both the particulars of digital practice and its broader philosophical, aesthetic, and historical implications, *Meta-Morphing* takes up and grasps the complexity of the morph along two complexly interrelated and interdependent axes: one formal and technical and the other historical and cultural. Neither one alone would do the figure justice.

Notes

1. In relation not only to morphing but also to computergraphics in general, an entire essay might well be written on the latter's mobilization of the word "render" and its variants. From the Latin *reddere* (to give back), the word is rich in its entailment with digitization and the reduction of all input to a single and fundamental binary code—a sort of primal digital soup. According to the *Oxford English Dictionary,* to "render" means not only to violently split, divide, or tear, but also to restore, return, or "surrender." Other definitions include to make "something over to another"; "to reproduce or express in another language," to "translate"; to "reproduce, represent," and "depict," especially by "artistic means"; and to "return by reflection or repercussion." Of great relevance in the present context, "rendering" also refers to "the process of extracting, melting, or clarifying fat." This connection of rendering with clarifying would seem to evoke Western notions of fixing borders, categories, and identities, and yet here clarifying is a dissolution, a breaking down and melting. These confusions and contradictions are nowhere better dramatized than in *Terminator 2: Judgment Day* (1991) when both the original Terminator and the T-1000 are "clarified" at the film's end in a vat of molten steel. *All* the meanings of "render" apply here. For more on this issue and the film, see Albert Liu, "Theses on the Metalmorph," *Lusitania* 1, no. 4 (1992): 130–42.

2. A particularly fine reading of the political implications of the morph can be found in Doran Larson, "Machine as Messiah: Cyborgs, Morphs, and the American Body Politic," *Cinema Journal* 36, no. 4 (1997): 57–75. I am indebted to this essay for bringing to my attention the passage from Vico that serves as this introduction's epigraph.

3. Michel Foucault, *The Archaeology of Knowledge,* trans. A. M. Sheridan Smith (New York: Harper and Row, 1972), 17.

4. Cecelia Tichi, *Shifting Gears: Technology, Literature, Culture in Modernist America* (Chapel Hill: University of North Carolina Press, 1987), 240.

5. I am most grateful to E. Ann Kaplan for raising the connection between my articulation of the morph as a "meta" object and metaphor. Of interest here, once

again, are the strange connections and contradictions revealed by etymological grazing. On the one hand, the *OED* defines "meta" as meaning "across" and suggesting a "transfer" from one context to another and a "substitution"—hence its status as prefix in "metaphor," where it qualifies "phor" (taken from the Greek *morphé,* itself from *pherein,* meaning "to bear"). However, and of particular relevance to the word "metamorphosis," "meta" is also defined as "denoting a nature of a *higher* order" and of an order of a "more *fundamental* kind" (italics mine). This seeming contradiction and conflation does, in fact, define the dual nature of the morph as at once a higher-order metaphysical construct and an elemental physical materiality.

6. Three of the four panelists have contributed essays to this volume based on their earlier presentation (Matthew Solomon, Kevin Fisher, and Victoria Duckett). Unable to contribute because of other commitments, Ilsa J. Bick, M.D., the fourth panelist, presented a most interesting, psychoanalytically focused paper (whose absence here I greatly regret) entitled "'My Vicarious Depravity': A Developmental Appraisal of Transmogrification in Film."

7. Stephen Prince, "True Lies: Perceptual Realism, Digital Images, and Film Theory," *Film Quarterly* 49, no. 3 (spring 1996): 17, 36.

8. Much has been made of the lithe and graceful T-1000, particularly as it has been seen as a "feminized" counterpart to Schwarzenegger's muscle-bound and quasi-mechanical cyborg. Indeed, throughout much of the writing on morphing, the morph's general fluidity has been dealt with in relation to gender bifurcation. Although these readings (whether for or against) are persuasive, there is a certain lack of significant attention paid to other meanings attached to slightness of figure, fluidity, flow, and grace that actually would conflict with certain connotations of the feminine (whereby the "mysterious" and "chaotic" is also characterized as materially immanent, as—if you will—a concrete site of "resistance" to the transcendent and transcendental). Most of those aspects of the T-1000 often read as "feminized" can also be read as being about overcoming immanence, bodily thickness, and resistance. In this, its utopian imagination is securely attached to (white, male, patriarchal) Cartesian philosophy.

I. Metamorphosis, Magic, and Mythology

Figure 1.1 The acknowledged master of chapeaugraphy was the French conjurer and shadowgraphist Felicien Trewey. Courtesy of the Houdini Collection, Harry Ransom Humanities Research Center, University of Texas at Austin.

1

Quick-Change "Twenty-Five Heads under One Hat" in the 1890s

MATTHEW SOLOMON

With its technological capacity to project the illusion of movement across a sequence of still images, the cinema introduced a new form of visual transformation at the end of the nineteenth century. Yet it is important to recognize that the new medium appeared against the backdrop of a theatrical culture itself characterized by transformations of various kinds. During the 1890s, cinema was often seen alongside, or as a part of, performances by professional illusionists, who were a staple of the variety theater's diversified programs and who sometimes presented moving pictures as a featured marvel of magic shows. In addition to the conjuring tricks and stage illusions with which magicians are most closely associated, a number of turn-of-the-century illusionists specialized in adeptly performed transformations such as quick-change artistry, the rapid alteration of character through costume changes; chapeaugraphy, the manipulation of a piece of felt to form different hats; and shadowgraphy, the use of the hands to create human and animal figures in a beam of light. Such practices were independent of, but not entirely differentiated from, the realm of the professional magician. Nearly forgotten and seldom seen today, quick-change, chapeaugraphy, and shadowgraphy each relied on physical dexterity, in conjunction with particular accoutrements, to create striking transformations of character. Localized around the body of the performer, these metamorphic practices, when presented by such skilled performers as Leopoldo Fregoli, Felicien Trewey, and David Devant, inspired an awe in enthusiastic audiences that rivaled—and often seems to have exceeded—the wonders that cinema and other new technologies could offer.

The field of "pre-cinema" has generally been limited to the technologies and procedures that seem directly to prefigure the cinematic apparatus of film camera and projector. Thus serial photography and magic lantern projections often figure centrally in accounts of the prehistory of cinema. Practices such as quick-change, chapeaugraphy, and shadowgraphy do not. Although a number of pre-1900 films depict these practices in various forms, the cinema's appropriation of transformative performance was ultimately quite brief, and performed metamorphoses seem to have been largely superseded by technological transformations in film not long after the turn of the century. Charles Musser contends that the introduction of projected motion pictures in the mid-1890s is best viewed not as the invention of a new medium but as one of several significant shifts in the history of "screen practice": "In such a history, cinema appears as a continuation and transformation of magic-lantern traditions. . . . It is this sense of continuity that must be reasserted if we are to understand [historical] transformation as a dialectical process."[1] Treating quick-change, chapeaugraphy, and shadowgraphy as "pre-cinematic" highlights a number of historical discontinuities, since these practices, unlike those of the magic lantern (such as the dissolve), were never really absorbed into the cinematic repertoire. Placing metamorphic performance within a longer history of transformation that includes not only the emergence of cinema but also the contemporary proliferation of digital media, however, foregrounds a significant set of continuities. Viewed from the late twentieth century, one hundred years later, when the cinematic is being increasingly replaced by the digital, quick-change, chapeaugraphy, and shadowgraphy take on an added significance, appearing not so much archaic as visionary.

One of the first published accounts of the cinema, from 1896, begins with a commentary on the name given to the apparatus. The author, claiming that "its long name kept me away from the new invention," critiques the word *cinématographe* before describing the characteristics of the "wonderful invention":

> Although unwilling to quarrel with William Shakspere about his statement that the rose would smell as sweet by any other name, I can't help thinking that "Cinématographe" is a nasty word for busy people. It has

a terrifying effect upon the man in the street who calls an entertainment a "show."[2]

While this unsigned account of "Lumière's five-syllabled invention" continues to juxtapose cinema with other recent inventions such as the phonograph and telephone, the foregoing excursus on nomenclature suggests a linguistic comparison between "cinematography" and similar polysyllabic words such as "chapeaugraphy" and "shadowgraphy" that were coined to designate *non*technological modes of inscription at the end of the nineteenth century. During the 1890s, "cinematography" was one of a succession of neologisms that included "telegraphy," "phonography," and "radiography," but its "writing in movement" suggested comparisons not only with other technologies of representation and communication but also with contemporaneous performances of chapeaugraphy, "writing in hats," and shadowgraphy, "writing in shadows." Reconstructing these latter, nearly obsolete forms of entertainment and their reception allows us to reinsert cinema into the cultural—as well as the linguistic—context of other performative modes of *écriture.*

Practices such as shadowgraphy, chapeaugraphy, and quick-change made up part of an extensive performance culture emphasizing transformation, a culture that existed alongside the cinema into the second decade of the twentieth century. This theatrical culture was an important referent for the early cinema, especially before 1900, as demonstrated by a number of early films that depict and modify these metamorphic practices. A number of the magical films that show quick-change, chapeaugraphy, and shadowgraphy feature performances by Fregoli, Trewey, and Devant, who, as internationally known masters of their respective arts, were at the forefront of a relatively small cadre of virtuoso performers during the 1890s. Each, it should be noted, in addition to appearing in films, played relatively important roles in film production and exhibition before the turn of the century.[3]

Quick-change, or protean, artists are the clearest example—and perhaps the epitome—of the nineteenth-century theatrical tradition of transformational performance. They performed as part of variety programs and, in certain cases, in full-length stage shows. In both protean sketches, where the changes were framed within a narrative playlet, and straightforward quick-change acts, the performer rapidly assumed a series of widely different personae that varied wildly in gender, age, social class, and occupation. Much of the appeal of protean artistry hinged on the speed with which an individual performer could make costume changes that effected complete and striking transformations of character.

Leopoldo Fregoli was probably the best-known and most acclaimed protean, or quick-change, artist at the end of the nineteenth century. His amazing ability to make numerous, seemingly instantaneous changes of character through rapid alterations of his physical appearance, voice, and comportment set him apart from other quick-change artists. In 1900 one reviewer "gladly called [Fregoli] the most striking of the Protean men, past, present and future," stressing the "lightning speed" of Fregoli's transformations, as well as the surprises and humor of his show:

> During three hours. . . . Fregoli sufficed all connoisseurs, satisfied the most hard-to-please, as with him, one jumps from surprise to surprise and the "Ah!"s of admiration are followed by the "Oh!"s of surprise. . . . He is simply the most astonishing protean artist that could be. . . . Quick as lightning. . . . Fregoli goes out and reenters, changes costume, . . . changes sex. He must vary his voice from deep and low bass . . . to the opposite . . . a shrill soprano, so ventriloquism comes to his aid. . . . His grand opera in a prologue, one act, four tableaus and an epilogue is well the most ridiculous and the most hilarious thing that could appear on the boards of a theater.[4]

Fregoli's astounding vocal range (soprano, contralto, tenor, baritone), his fluency in several different languages (including Italian, French, English, and Spanish), and his ventriloquial skills allowed for the surprising auditory transformations that were a crucial part of his performance.

Unlike later quick-change artists who made changes behind transparent sets or onstage, Fregoli made the costume and hairpiece changes that filled out the appearances of each of his characters just out of view of the audience, in about three seconds, with the help of several assistants. Fregoli himself claimed the most difficult aspect of his act was the necessary de-synchronization of gesture and voice—continuing to speak in a calm voice to the audience in the auditorium while he hurried madly to change costumes just out of view. While using speech to bridge character transformations, many of Fregoli's metamorphoses also relied on what he termed a "reverse step." Fregoli would exit through a doorway, then appear to immediately step back from the doorway so that his back was visible to the audience. In fact, a similarly dressed assistant had taken his place so that Fregoli could change costumes and reenter the stage from another entrance. The great multitude of stage personae Fregoli would adopt in his performances is suggested not only by report of his eight hundred costumes but also by his epic boast that he could cover the base of the Eiffel Tower by spreading out his collection of twelve hundred wigs side by side.[5]

Fregoli was contrasted with other quick-change performers because his virtuoso talents of impersonation set him apart from others who merely made costume changes with dexterity:

> In fact, two things are required from the quick-change artist, dexterity in movement and skill in character-acting and improvisation. The marvellous mechanic may be the dullest of entertainers, and vice versa. . . . Fregoli . . . is mechanically marvellous and artistically proficient.[6]

As a performer, then, Fregoli was much more than a "mechanical marvel" because his "skill in character-acting" allowed him to assume the complete personae, as well as the habiliments, of the many different characters into whose costumes he rapidly slid. The truly entertaining quick-change artist created successful character impersonations through nuanced performance, for clothes never make the man or the woman, except in a strictly "mechanical" sense.

By the second half of the 1890s, Fregoli's shows were typically divided into two separate parts, one of which was narrative, and the other non-narrative. During the first part of the show, Fregoli would present one of his protean playlets, some of which involved a dozen or more characters, in which he would, of course, take all the roles. The second part of the show was invariably taken up by "Eldorado," an "action-comic-mimic-lyric-dramatic-musical with about sixty transformations," in which Fregoli impersonated the appearances, voices, and performance styles of a multitude of well-known theatrical personalities. "Eldorado," or "Paris-Concert," was presented as a faux vaudeville show in which Fregoli played the roles of all the performers, often at the end of intermission after he announced that all the other artists had failed to appear. The variety show included several different singers (both male and female), ventriloquists, magicians, well-known conductors, as well as imitations of Loïe Fuller's serpentine dance, a parody of "Trilby," and a number of other variety turns.[7] A 1900 account explains:

> His gusto does not stop for one second . . . what interior flame, what springs of steel in his nimble, active, jittery legs, he gives evidence of in these various sketches which he entitles the "Eldorado," . . . which resembles nothing that one has ever seen, for one feels Fregoli is always the

> artist. This gumby of the café-concert is a marvel of imitation. . . . Quitting the stage, he descends into the orchestra to create illustrious heads in the midst of the musicians, and with time to fix a wig, here it is with striking exactness Rubinstein, president Krüger, general Joubert and Garibaldi. Next he conducts the orchestra, in a tour of the celebrated musicians, imitating their tics and mannerisms in beating time. Rossini, Donizetti, Verdi, Wagner, Meyerbeer, Mascagni, Gounod, Bizet, Ambroise Thomas. For the finish, Fregoli performs the serpentine dance to "loïefulleresque" perfection; finally, this man, who during the evening indulges in sixty or eighty transformations, gives you the unbelievable illusion of conjuring himself and of reappearing immediately before you in black clothes.[8]

Beginning in 1896, Fregoli sometimes concluded the show with the exhibition of films featuring himself.[9] During these screenings, Fregoli would add synchronized voice and song ventriloquially to the screened images depicted by his "Fregoligraph."[10]

One of the most popular of Fregoli's protean sketches was "The Chameleon," which he had begun performing in the 1880s. He called it "a 'tragic-dramatic-comical' comedy in one act" and often embellished the roles of the husband, his unfaithful wife, and her lover by interpreting them in the acting styles of famous actors.[11] Because he could mutate into any one of the characters without warning, Fregoli's rapid transformations—which were sometimes accomplished in conjunction with other techniques of stage illusion—were nevertheless relatively unpredictable, as suggested by the following description:

> He would exit the stage on the right dressed as a vaporous young woman in a ruffled dress and umbrella. He would return almost instantaneously on the left, with the rhythm of a cake-walk: and there would be a suave fellow with monocle following the young miss. Then one would see him again dressed as a young girl and immediately afterwards, again dressed as the fellow. But that did not suffice him: the young miss would enter once again. She would sit reclined on a bank and open her umbrella with affectation. Just as soon, without it seeming so, Fregoli would abandon the dress and the umbrella—which continued to suggest a young girl—descend through a trap-door and would reappear as the fellow. He would approach the young miss, his hand over his heart, and declare his love to her. She would protest (Fregoli still speaking). She would call

for help. The authoritarian voice of the guardian of the peace would respond; the gallant chap would flee, startled. And there would be Fregoli dressed as the town sergeant.[12]

In his performances, the difficult work of the quick-change took place out of the view of the audience. Yet Fregoli's character transformations still provoked surprise due to their rapidity, their unpredictability, and his consummate skills of impersonation. Unlike "Eldorado" and analogous displays of quick-change, in which costume changes were congruent with transitions between different self-contained presentations, the transformations in Fregoli's protean sketches were, in a sense, embedded in the story. Performed skillfully, protean playlets relied on a symbiosis between narrative and metamorphosis.

In contrast, chapeaugraphy placed the means of transformation in the spectator's field of vision, distilling character impersonation to manipulations of a single piece of apparel, the hat, combined with corresponding alterations of the performer's physiognomy. Accomplished chapeaugraphers, unlike most quick-change artists, allowed discerning spectators to attend to the manner in which the transformation was accomplished by making the manipulations of the "chapeau" in view of the audience. The acknowledged master of chapeaugraphy was the French conjurer and shadowgraphist Felicien Trewey. Trewey called his chapeaugraphy act "Tabarin, or Twenty-Five Heads under One Hat," after Antoine Girard Tabarin, the legendary early-seventeenth-century French clown and farce player known for "a floppy felt hat capable of any number of metamorphoses for comic effect."[13] In performance vignettes on the Pont Neuf of Paris, Tabarin had used a hat of ambiguous shape that could assume various forms as a transformative accessory in burlesque impersonations. Tabarin called his chapeau "true raw material, indifferent to all forms . . . having almost no essential form except formlessness . . . as this hat strikes by great alterations, for there is no moment, no instant, when it does not receive a new figure."[14] During the 1890s, performances of chapeaugraphy by Trewey and his imitators involved rapid contortions of a stiff but flexible ring of felt (shaped like a wide-brimmed hat without a crown) to (trans)form a number of different types of headgear.

Like Fregoli, Trewey imitated popular celebrities and recognizable character types, but his mimicry was accomplished with fewer markers of identity. Like Fregoli's "Eldorado," Trewey's "Tabarin" was essentially nonnarrative, presented as a sequence of twenty-five different portraits of individuals in relatively quick succession. One contemporaneous observer, Henry Ridgely Evans, describes the "Tabarin" as performed by Trewey:

> Thanks to a piece of black felt cloth, circular in shape, with a hole cut in the center, Trewey is able to manufacture in a few minutes all of the varieties of headgear required for the Tabarin. For example: Napoleon—A couple of twists of the cloth, and lo! you have a representation of *le chapeau de Marengo,* the little cocked hat which Napoleon made famous, and about which so many legends cluster. With this hastily improvised hat on his head, Trewey assumes the Napoleonic attitude—one hand thrust into his vest, the other behind his back. His physiognomy is that of the great Emperor, as depicted by the painters of the Imperial régime. The likeness is perfect. And so with fat French priests, soldiers, bonnes, landladies, artists, diplomats, etc. It is a portrait gallery of French types.[15]

The same observer claimed that Trewey's true skill was not his practice and ingenuity with the felt ring but rather the "mobile features" of his face:

> With the brim of an old felt sombrero, Trewey is able, by dexterous manipulation, to construct every variety of headgear. . . . It is not these varieties of headgear that astonish the audience, but Trewey's facial interpretations of the different types of character assumed. . . . It is a facial pantomime of exceeding skill.[16]

Unlike other less-skilled chapeaugraphers, who were unable to "command the requisite mobility of features" and required "the use of certain accessories" (false mustaches, beards, pipes, and other items) for their characterizations, Trewey "dispensed with all kinds of 'make up,' and relied entirely upon his marvellous mobility of features and the 'ring.'"[17]

In chapeaugraphy, impersonation was in large part defined by a metamorphosing piece of headgear. Framing the face, the hat is a primary focal point for visual interest and, more than any other article of clothing, possesses a uniquely disproportionate power of expression. Donning the deftly manipulated "chapeau," the performer could thus assume the guise of a sailor, cowboy, priest, toreador, clown, nun, admiral, jockey, devil, or witch—to name but a few of the characters—with a few swift twists of the felt ring.[18] Chapeaugraphy highlighted the immediate signifying potential of a number of different forms of headgear for turn-of-the-century spectators, often drawing on existing stereotypes of ethnic and national identity. It was, Evans says, "an international portrait gallery, and we see represented in the 'Tabarin' Irishmen, Scotchmen, Englishmen, Chinamen, and other nationalities."[19]

Shadowgraphy, as it was presented by skilled performers during the 1890s, involved mimicry rather than impersonation. As a theatrical entertainment, the marvel of shadowgraphy (or *ombromanie*) was the ability of performers to configure their hands adroitly in a beam of light so as to create nuanced shadow profiles of people and animals highly praised for their lifelike detail and movement. Different from shadow puppetry, shadowgraphy involved only the carefully configured hands of the performer—although parts of the arms and shoulders, along with small shapes of card, were sometimes used to complete the profiles, and at least one performer, Chassino, used his feet—rather than flat shadow puppets placed behind a screen. In the theater, the shadowgraphist sat or stood between a circular point source of light and a screen on which the shadows appeared; both rear and front projection were employed, depending on the individual performer and the depth of the stage.[20]

In 1895 the magic journal *Mahatma* commented on "what practice will do when combined with grace" in shadowgraphy "to make an entertainment that cannot be surpassed":

> To the casual observer Shadowgraphy seems an easy feat to learn, but in actual practice and when well done there is nothing more difficult than shadowgraphs, requiring as it does nimble fingers and the patience that but few performers possess. There are but few men of note in this country to-day doing this act . . . for, it not only requires great skill but an ingenious brain. . . . The skillful performer should be seen to be appreciated.[21]

Two of the most accomplished practitioners of shadowgraphy during the 1890s, Trewey and English magician David Devant, presented numerous different shadow pictures, imbuing each with lifelike movement. In William J. Hilliar's account, "Shadowgraphy in his [Trewey's] hands becomes a grand art . . . not . . . so much in his novel figures . . . as in his marvelous ability to make each and every little detail appear to be absolutely life-like."[22] Evans adds:

> Trewey has made his hands so supple that he can not only form the most diverse figures upon a screen, but can also give them motion and life. The swan smoothing its plumage, the bird taking flight, the cat making

its toilet, the tight-rope dancer, who, after saluting the public, rubs chalk on her feet before walking on the rope, are true wonders, and it is hard to believe that these perfectly accurate profiles are obtained solely by means of the shadow of the hands.[23]

The articulation, movement, and detail of the screen images belied the apparent simplicity of the body parts used to create them. One of the wonders of shadowgraphy was the way that performers could cast lifelike shadow likenesses with contortions of the fingers.[24]

As well as enacting amusing shadowgraphic pantomime sketches with his hands, "Trewey always made a specialty in his performances of producing finger silhouettes of celebrated people."[25] Trewey's "international portrait-gallery of shadows" included shadow profiles of Gladstone, Czar Alexander III, Bismarck, Lord Salisbury, Zola, and President McKinley. One observer noted, "For each and every one of them is designed a certain marvellously appropriate movement; and even the great personages whose portraits appear on the disc are made to exhibit some mannerism or characteristic whereby they are known."[26]

In addition to being able to impart shadow portraits with lifelike movement, skilled shadowgraphists, unlike drawing-room amateurs, could seamlessly transform one profile into another by reconfiguring their fingers in front of the lamp. Indeed, as one shadowgraphist noted, "The human hand, the most perfect of instruments, possesses in a most extraordinary degree the faculty of being able to be brought into . . . an endless, rapid, and surprising number of changes."[27] An 1897 account describes such metamorphoses: "Many of these portraits are transformation portraits, one changing into another in sight of the audience, but yet not so quickly that the various motions are indistinct, or untraceable by the keen-eyed."[28] Referring to such transformation portraits as "transforming finger photos" because of their apparent realism, the author describes how Devant also emphasized the manual procedure he used to create deceptively lifelike shadows:

As Mr. Devant's hands enter the illuminated disc they are quite separate, all the fingers being extended. The operator then proceeds dexterously to "mould" his subject, but in such a manner that all may behold the clever evolution of the figure. The placing of the hands and the disposition of each figure are swiftly seen by the intelligent audience, who appreciate this method far more than they would the instantaneous appearance of perfect figures.[29]

Metamorphic transformations between shadow portraits foregrounded an inherent tension between the performative constructedness and the representational verisimilitude of the shadowgraph. As traced by Johann Casper Lavater's profile-machine beginning at the end of the eighteenth century, the silhouette itself had seemed to record a two-dimensional likeness of the subject—although, unlike photography or portraiture, implying that the distinctive features of the individual could be found in an oblique view. Skillfully presented shadowgraphy, however, destabilized the apparent representational accuracy of the silhouette through manual dexterity. Transformation portraits, like quick-change artistry and chapeaugraphy, made human physiognomy, the perceived guarantor of individuality, into a constantly mutating function of performance.

The virtuoso shadowgraph performance, therefore, did not efface the performer positioned behind the screen in order to intensify the projected fictions but rather emphasized the performative nature of the screen images in order to continually remind the spectator of the shadowgraphist's very presence. As Devant noted in his advice to prospective shadowgraphists, "the average audience will be as anxious to watch your hands as to see the shadows."[30] Like other contemporaneous "lightning entertainments," shadowgraphy and chapeaugraphy centered on the hands of the performer as the means of metamorphosis.[31]

Part of the appeal of shadowgraphy, chapeaugraphy, and other such practices was probably related to the way that these entertainments, like magic tricks, foregrounded what Neil Harris terms the "operational aesthetic." Although Harris locates an "aesthetic of the operational, a delight in observing process and examining for literal truth,"[32] most clearly in the nineteenth-century popular curiosity that focused on the workings of new technologies, one can see it manifest in modes of metamorphic performance that tempt the viewer to discover the methods of transformation. However, attempts at demystification like books of shadowgraphs fail to provide the sought-after revelation, for such guides illustrate only the proper positions of the hands and fingers—the means of transforming the shadows (as well as of animating the figures) remains inscrutable. Indeed, a booklet that was sold at Trewey's performances suggests the contra-technological attraction of performative metamorphic practices and the impossibility of their revelation:

> Those who see his entertainment cannot fail to observe that the merit of the performance lies in the marvellous skill which it demands. One sees

> many clever conjurers, men who produce strange and apparently impossible results by some inexplicable means. It is wonderful, but the audience knows always that the strange effect is only produced by a deception of some sort, and that the conjurer has only succeeded in pretending cleverly to do what was only apparently impossible. The ordinary apparatus conjurer excites for this reason only our wonderment and not our admiration. His performance has no merit at all when one sees "how it is done." Trewey's entertainment is of a different stamp. There are no elaborate apparatus or concealed mechanical contrivances, no false bottomed and double-lidded boxes. His "properties" are of the simplest kind, and his performances excite one's amazement on account solely of his skill. He really does what, to everyone who sees it, is absolutely impossible, and which is only accomplished by himself by reason of his own cleverness and of a phenomenal development of manual dexterity.[33]

Unlike apparatus magic and stage illusions, the wonder of shadowgraphy and chapeaugraphy was not based on deception, nor entirely on the operational aesthetic. Rather, it was a pure function of extreme dexterity and virtuoso performance and, as such, could not be exposed.

Harris's notion of the operational aesthetic, which hinges on the possibility of real revelation, highlights the contrast that Sergei Eisenstein makes between the "trick" and the "attraction" in popular theater: "The trick . . . is a finished achievement of a particular kind of mastery . . . absolute and complete *within itself* . . . the direct opposite of the attraction, which is based exclusively on something relative, the reactions of the audience."[34] Following Eisenstein, a greater potential for revelation makes a "trick" into more of an "attraction," although the two are rarely mutually exclusive. For example, acrobatics, which Eisenstein cites as a "trick" (or "stunt") "for the most part," contain no such potential. In the case of performed transformations, the contrast is suggested by the different relations of the audience to, on the one hand, Trewey's chapeaugraphy (a trick of manual dexterity) and, on the other, to Fregoli's quick-change (an attraction that hides its method from view, implying the reactions of an audience). If, as Harris asserts, technology is the definitive locus for the operational aesthetic, then the replacement of the performer by a technological apparatus makes practices such as quick-change, chapeaugraphy, and shadowgraphy even more attraction-like.

Different combinations of performative and technological modes of transformation can be seen in several films made before 1900 that present originally performative practices such as chapeaugraphy, shadowgraphy, and quick-change as cinematic spectacle. The Lumières' *Chapeaux à transformation,* a "comic view," suggests a short performance of chapeaugraphy but shows a performer donning six different hats (along with masks, wigs, false mustaches, and beards) in succession rather than the manipulation of a flexible "chapeau." In contrast, *Devant's Hand Shadows* simply used the cinematic apparatus to show "some of Mr. Devant's most original and popular ideas," the catalog description promising that the film "is so done as to give the same effect on the screen as if the performer himself were at work."[35] The latter film, grouped under the rubric of "Conjuring, Acrobatic, and Stage Performances by Well-Known Artistes" with other performative wonders including "The Deonzo Brothers, in Their Wonderful Tub Jumping Act," an "Acrobatic Performance, by Sells & Young," and "Mr. Maskelyne (of the Egyptian Hall), Spinning Plates and Basins" in a 1901 catalog, highlighted the singular physical talents of the performer. Similarly, in a motion picture of "Cronin, American Club Manipulator," "The dexterous way in which ordinary Indian Clubs can be made to execute apparently impossible movements is well shown. They appear to be endowed with life, and obey the will of the trained manipulator."[36] Méliès's *Transformations éclair,* on the other hand, created a marvelous effect that was entirely different from that of the performed quick-change by substituting the transformations of film editing for those of the performer.[37] The catalog description boasts of a revelation that the film does not make: "A man makes twenty complete character changes in two minutes, combining with them dances. The changes are made in full sight of the audience. Blondi, Fregoli and Mons change behind the scenes."[38] The changes, although made in full sight of the audience through the substitution splice, shift the transformations from the register of performance to technology, withholding the promised revelation of the trick and replacing it with a uniquely cinematic attraction.

During the first decade of film history, protean performers were gradually eliminated from cinematic spectacles. This historical shift runs parallel to what Donald Crafton describes as the "evolutionary change from the 'lightning cartoonist' films to the true animated cartoon" shortly after Emile Cohl's film *Fantasmagorie* (1908): "There was a noticeable shift of emphasis from the performer to the drawings. . . . The drawings are now of primary importance and the hand is secondary—a mere vestige of the old vaudeville tradition that would disappear."[39] The elimination of the dexterous hands of virtuoso shadowgraphists, chapeaugraphers, and quick-change artists by film technology would produce the genre of the trick film, but something significant was lost in this shift that the preceding discussion has only hinted at. The loss is perhaps best described by Trewey, who like Fregoli and Devant occupied an ambivalent position between transformative performance and film at the turn of the century. As a close friend of Antoine Lumière, Trewey had been chosen to introduce the *cinématographe* to London as well as at a number of other theatrical premieres in 1896, but he left film work within a year or two.[40] Interviewed in 1912, Trewey commented: "The machines and the films have improved wonderfully, but not the way in which the films are projected. The projectors have improved, and they now use electric motors to run them, but an electric motor is a machine, and a machine can never equal a man at the handle."[41]

Employed as the very motor enabling and regulating projection, or in conjunction with cinematic technology as in films depicting "lightning entertainments," the human hand could retain some of the power it had so ostentatiously shown off in metamorphic performances of the 1890s. As one author asserted in the beginning of a 1913 shadowgraphy book:

> Every part of the human frame is a storehouse of wonder to the student, and it is often left to the makers of amusement to reveal its extreme possibilities; the gymnast, the contortionist, the "strong man," the juggler, and many others of their class, each shew the marvellous degrees of skill, endurance, and adaptability to which various functions of the body can be trained.[42]

Unlike the jarring transformations of montage, performed transformations had the potential for a fluidity that is comparable to what Eisenstein characterizes as "plasmaticness," a notion that he formulated with respect to the animated film. Eisenstein characterizes plasmaticness as the "rejection of once-and-forever allotted form, freedom from ossification, the ability to dynamically assume any form"—reminding one of Tabarin's description of his polymorphous hat. For Eisenstein, plasmaticness crystallizes around the myth of Proteus, the son of Poseidon, god of the sea, who in Greek mythology had the ability to change his shape at will as well as to foretell the future:

> A being of a definite form, a being which has attained a definite appearance and which behaves like the primal protoplasm, not yet possessing a

> "stable" form, but capable of assuming any form and which, skipping along the rungs of the evolutionary ladder, attaches itself to any and all forms of animal existence.... One could call this the *protean element*, for the myth of Proteus (behind whom there seems to be some especially versatile actor)—or more precisely, the appeal of this myth—is based, of course, upon the omnipotence of plasma, which contains in "liquid" form all possibilities of future species and forms.[43]

Practices like protean artistry and chapeaugraphy sought the fluidity and malleability of their namesakes, the mythic Proteus and the quasi-mythic hat of Tabarin, as well as the capacity for rapid metamorphosis that shadowgraphy seems to have made possible in two-dimensional monotone representations. Although the plasmatic potential of metamorphic entertainers presented an exciting set of possibilities for the emergent cinema, the transformative practices that they had perfected during the 1890s quickly came to represent the path not taken by the new medium.

Largely abandoned by cinema except in certain animated films, this path reappears with digital media, which significantly expand the capabilities of what Vivian Sobchack describes as the "film's body."[44] Digital manipulations allow color film, with its verisimilitude and illusion of three dimensions, to assume the fluid state of plasma, challenging cinema's basic ontology and returning it to a seemingly previous state. Although the materiality of performance has been effaced, the uses to which digital transformation are being put are not so different from that which performers of the 1890s sought to create through physical dexterity. The myth of *T2* is truly much the same as the myth of Trewey, the myth of Tabarin, and—ultimately—the myth of Proteus.

Notes

An earlier version of this essay was presented at the 1996 Society for Cinema Studies Conference in Dallas as part of the "Meta-Morphing" panel chaired by Vivian Sobchack. The author wishes to acknowledge the helpful comments and suggestions offered by Vivian Sobchack, Yuri Tsivian, and Peter Wollen during the research and (re)writing of the essay.

1. Charles Musser, *The Emergence of Cinema: The American Screen to 1907* (Berkeley: University of California Press, 1990), 15.

2. "A Wonderful Invention: The Cinématographe of M. Lumière," *Sketch* (London), 18 March 1896, 323.

3. Erik Barnouw, *The Magician and the Cinema* (New York: Oxford University Press, 1981), 50–58, 62–65.

4. Dom Blasius [Henri Rochefort, pseud.], "Premières Representations: Fregoli dans ses transformations," *L'Intransigeant,* 22 January 1900, 2; translation mine. See also Jean Nohain and François Caradec, *Fregoli 1867–1936: Sa vie et ses secrets* (Editions de la Jeune Parque, 1968), 14–16; and Patrick Rambaud, *Les microbolantes aventures de Fregoli racontées d'après ses Mémoires et des témoins* (Paris: Éditions François Bourin, 1991), 119–20.

5. Fregoli, quoted in Nohain and Caradec, 65–67.

6. "Alhambra Theatre," *Daily Telegraph* (London), 10 March 1897, 4.

7. Nohain and Caradec, 65–76; and Rambaud, 91–92.

8. Blasius, 2.

9. Nohain and Caradec, 81–88; Rambaud, 140–43; and Georges Sadoul, *Les Pionniers du Cinéma (de Méliès à Pathé) 1897–1909* (Paris: Éditions Denoël, 1947), 46. It is unclear whether these films revealed the actual methods Fregoli used to change costumes offstage. Adolph Zink, a midget who appeared on the vaudeville stage, combined performed quick-change with motion pictures, doing "imitations of Edna May, May Irwin, etc., etc., making changes back of the scenes while the biograph sheet is dropped between each change, and moving picture views of his making the changes are shown on the sheet." H. A. Daniels, "Criticism of Keith's Theatre, Phila., Pa., Nov. 9th (1903)," *Managers' Report Book* 1, 62, Keith/Albee Collection, Special Collections, University of Iowa.

10. Glauco Pellegrini, "Fregoli ou Le premier 'appareil' de projection sonore," *La Revue du Cinéma,* n.s., no. 14 (June 1948): 48–51.

11. Rambaud, 53–76; and Nohain and Caradec, 35, 70–71.

12. Pellegrini, 49; trans. Tami Williams.

13. Donald Roy, "Tabarin," in *The Cambridge Guide to World Theatre,* ed. Martin Banham (Cambridge: Cambridge University Press, 1990), 940.

14. Tabarin, *Les Oeuvres de Tabarin,* ed. Georges d'Harmonville [Paul Lacroix, pseud.] (Paris: Libraire Garnier Frères, n.d.), 131; translation mine. See also "Les Fantaisies Plaisantes et Facetieuses du Chappeau à Tabarin," 255–60.

15. Henry Ridgely Evans, *The Old and the New Magic* (Chicago: Open Court, 1906), 341.

16. Henry Ridgely Evans, "Shadowgraphy," in *Magic: Stage Illusions and Scientific Diversions,* ed. Albert A. Hopkins (London: Sampson Low, Marston, 1897), 179.

17. Hercat [R. D. Chater, pseud.], *Chapeaugraphy, Shadowgraphy, and Paper-Folding* (London: Dean and Son, 1909), 8.

18. Ibid., 7–31.

19. Evans, "Shadowgraphy," 179.

20. Alber [pseud.], *Les Théâtres d'Ombres Chinoises* (Paris: E. Mazo, 1896), 123–39; and Denis Bordat and Francis Boucrat, *Les Théatres d'Ombres: Histoire et Techniques* (Paris: L'Arche, 1956), 130–49.

21. "Shadowgraphy," *Mahatma* 1, no. 2 (April 1895): 5.

22. William J. Hilliar, *Modern Magicians' Hand Book: An Up-to-Date Treatise on the Art of Conjuring* (Chicago: Frederick J. Drake, 1902), 412.

23. Evans, "Shadowgraphy," 175. See also Henry Ridgely Evans, "The Shadows of a Clever Pair of Hands," *Cosmopolitan* 27 (June 1899): 163–66.

24. With Trewey, Devant, and a few of their contemporaries, as one sees in both accounts of their performances and the manuals they published, shadowgraphy was transformed from an inert display of a number of human and animal silhouettes—as depicted in Henry Bursill's books of hand shadows of 1859 and 1860—into a dynamic performance of a series of lifelike animated vignettes. Henry Bursill, *Hand Shadows to Be Thrown upon the Wall* (1859; reprint, New York: Dover, 1967), and *More Hand Shadows to Be Thrown upon the Wall* (1860; reprint, New York: Dover, 1971).

25. Hilliar, 420–21.

26. Bernard Miller, "Hand Shadows," *Strand Magazine* 14 (December 1897): 629.

27. Imro Fox, "Shadowgraphs, or Shadows of the Hand," *Tricks* 1, no. 1 (1 June 1901): 11.

28. Miller, 628.

29. Ibid., 626.

30. David Devant, *Hand Shadows* (London: S. H. Bousfield, 1901), 6.

31. Clay modeling, quick-sketch drawing, and paper folding were other entertainments that mobilized the manual dexterity of the performer to create transformational representations in different media. See also J. F. Burrows, *The Lightning Artist: A Treatise on Pictures in Smoke, Rags, Sand, and Paper, and How to Make Them* (London: Hamley Brothers, n.d.).

32. Neil Harris, *Humbug: The Art of P. T. Barnum* (Chicago: University of Chicago Press, 1973), 79.

33. *How It Is Done* (Middlesbrough: Jordison, 1893), 4.

34. Sergei Eisenstein, "The Montage of Attractions," in *Writings, 1922–1934,* trans. and ed. Richard Taylor (London: British Film Institute, 1988), 34–35.

35. Robert W. Paul, *Catalogue of Paul's Animatographs and Films* (1901), n.p.

36. Ibid.

37. A significant element of the overall aesthetic of Méliès, it should be noted,

centers on the instantaneous transformations of costume and headgear accomplished in his films through the substitution splice.

38. *Complete Catalogue of Genuine and Original "Star" Films (Moving Pictures) Manufactured by Geo. Méliès of Paris* (1903), 19, in *Motion Pictures by American Producers and Distributors, 1894–1908* (microfilm), comp. Charles Musser (Fredericksburg, Md.: University Publications, 1984–1985), reel 4.

39. Donald Crafton, *Emile Cohl, Caricature, and Film* (Princeton: Princeton University Press, 1990), 138.

40. Jacques Rittaud-Hutinet, *Le cinéma des origines: Les frères Lumière et leurs operateurs* (Paris: Editions du Champ Vallon, 1985), 186–97.

41. Trewey, quoted in John Cher, "Who Is the Father of the Trade? Interview with M. Trewey," *Bioscope*, 17 October 1912, 187.

42. Louis Nikola, *Hand Shadows: The Complete Art of Shadowgraphy* (London: C. Arthur Pearson, 1913), 11.

43. Sergei Eisenstein, *Eisenstein on Disney*, ed. Jay Leyda, trans. Alan Upchurch (London: Methuen, 1988), 21, 64.

44. Vivian Sobchack, *The Address of the Eye: A Phenomenology of Film Experience* (Princeton: Princeton University Press, 1992).

2

A Brief Disappearing Act

Animation and Animorphs

NORMAN M. KLEIN

A chalk line transforms into a man's whiskered face. A hand reaches across the drawing, then erases and redraws the face, aging it, changing its sex, its race. Caricatures of blacks, of Jews, of women's naked thighs appear and dissolve, what was called "lightning hand," at the turn of the twentieth century.[1] In 1907 Stuart Blackton filmed his lightning-hand sketches,[2] as did Winsor McCay four years later.[3] The memory of lightning hand reappears in Otto Messmer's Felix the Cat cartoons of the twenties. Even as late as the forties, Ward Kimble and other Disney animators perform lightning hand as a racy burlesque for soldiers, where the line drawing turns into a naked woman.[4]

Along with chalk—or ink—any number of substances have been used to indicate the human hand tangibly interfering: finger paint on glass, shifting sand, or simply programmers using algorithms to make shapes shimmer without mass.[5] But the effect is essentially the same in hundreds of animated shorts. One substance transmutes into another, from line to protoplasm and back again. And with this transformation, gravity itself, or time, transforms as well.

As the transformation begins, a second phenomenon takes over. Gravity itself seems to disappear. Laws of nature collapse. But very soon, an alternative law takes over, an uncanny logic where events go out of scale: the sculpture of morphing. It is mercurial. Flesh, or metal, flows like water, as in the "morphing" effect initiated with *Willow* (1988) and *The*

Abyss (1989), and made standard after *Terminator 2: Judgment Day* (1991).

Compare this to the metamorphic gag structure in hand-drawn cartoons: here, the "mercurial" does not run in a single stream. The gags flow sporadically but suggest an alternate rule of order. The imaginary space is infinitely more uneven—more an architecture of disunities. However, metamorphosis is far more lyrical than a pyramid of gags. A much different effect is at play. We need to set up new terms to clarify this difference—first through an exploration of metamorphosis in animation frame by frame, including computergraphic morphing; then a case study of cartoon animation; and finally a look at what these terms suggest about our political culture at large.

First of all, let me separate the verb from the noun and shorten *animated metamorphosis* to *ani-morph.*[6] At least that begins to isolate some of the variables. *Ani-morphing* can be defined as an animated cycle where metamorphosis takes place—for example, a walk cycle where a creature changes species. The body and proportions will be exaggerated, with "extremes" on either end. But frames inside, called "in-betweens," stabilize the action, make the switch more convincing, and also balletic, rhythmic.

Let us imagine a midpoint inside this cycle, between the extremes—a *lapse* or *hesitation,* what I call an ani-morph. The shift is suddenly not very stable. For a few frames, the object—the body, in this case—does not look like what it was, or what it will be. The ani-morph is literally *between* the rest of the cycle.

The audience may catch a glimpse of the hand at work—not the hand itself, but the tactile fact of a drawing on paper. It can distance the effect, like noticing the string of a marionette. In modernist terms, that glimpse into praxis is a self-reflexive device. It highlights the craft of the animator, not unlike "gesture" in modern painting. Oskar Fischinger's abstract films are essentially ani-morphs as sensory rhythm. In animation, then, unlike live-action film as a rule, this ani-morph can be extended almost indefinitely.

The verb *ani-morphing,* the noun *ani-morph:* these two terms make it easier to describe how lightning hand operates as a narrative of transformation. However, I do not mean narrative as in plot points lifted from Ovid's tales. Ani-morphing does not, in itself, generate a dramatic structure; it is closer to a visual fable about colliding atmospheres (the god descending to earth; the storm raining indoors). It is condensed magic realism, snippets from Gogol's tales, Kafka's *Metamorphosis,* or Schulz's *Street of Crocodiles*—and, of course, the Quays' sumptuous adaptation of Schulz's novel, filled with ani-morphs as colliding atmosphere. Real meat from the butcher is handled by puppets. Screws and dust in the workshop turn in reverse, as if time were going backward.

In other words, not only does the body morph, but the air itself as well. Let us suppose that a puppet learns to breathe like a human being. Gradually, its neck opens into gills that evolve before your eyes into a trachea. Then imagine this mutation across the entire screen. Two atmospheres appear to collide; they swarm into each other, like oil through water. At first, the creature breathes the air provided artificially by the puppeteer, or by the stop motion animator. However, the controls restrict movement like an iron lung. Then the puppet escapes into the street.

How does the animator turn this kind of atmospheric palimpsest—these ani-morphic tropes—into story? In simplest terms, the background morphs. Imagine a car hopping from the earth's atmosphere to the moon, merely by skipping past the broken yellow line. The background goes from paper to elephant skin. Or it bears witness to the change from wood to paper, like the cross-hatching in a nineteenth-century wood engraving. Depth and mass change hands. The space itself is "speaking." The background behaves as if it were alive. It forces changes that are uncanny, chaotic, even animistic, while continuously revealing the paper and ink used by the animators.[7]

How do these contrasts add up to a form of story—as in conflict, reversal, surprise? For clues, we should review how animation is drawn. To add life to a drawing, the identity of line should be unstable, to imply movement, breathing. Therefore, in life drawing classes, animators are trained to "forget" simple body proportions. Instead, they study "implied mass," lines that show the weight shifting from one leg to another. The model's hip is distorted as he strains to hold pose. That may be distressed even more; the outstretched arm is lengthened. These exaggerated torsos might look very sensual or might merely suggest physical discomfort, the presence of time and gravity. But on film, the results will look fiercely energetic, particularly if clever gaps are slipped into the cycle—as in extremes or ani-morphs.

Then there is the power of erasure, yet another tool that is ani-morphic, when the drawing is "cleaned up." We look at an animator's sketch pad. To capture implied mass, there is a blizzard of lines. From these, the most "active" (distressed) will be selected. The rest will be erased. The result should leave negative space or mass, which amounts to yet another ani-morph, but this version is not chaotic; it suggests control, nonwaste, stability. And it implies mystery as well: a phantom presence, as if a hundred pounds were hidden.

The drawings are then flipped, to see how they move. Empty spots, or glitches between actions, may be kept blank. What does this do? It is not quite like chiaroscuro in live action; it may not deepen the atmosphere. It is more like a camera zooming on a Rembrandt portrait until we see the cracks in black paint. The viewer is forced back to the surface. Vividly, as if behind fierce sunlight, the figures seem to disappear, like a sense memory dissolving. The drawing becomes *allegorical* in Walter Benjamin's sense of allegory—a ruin. It is a dialectical emblem, where all that appears natural is simultaneously artificial. Therefore, "events" that generated the drawing "shrivel up" but leave an absence one can sense. And "craftsmanship" is revealed from beneath, "like the masonry in a building whose rendering has broken away."[8]

For animation, these empty spaces are handled as ani-morphs—many chipped buildings at once, many ruins attached to a background that seems almost limitless in the number of erasures it can contain. Animation is shape-shifting in multiple, from colliding atmospheres to dissolving ink—phantom limbs. We imagine the shadow where a picture frame once hung.

All these fragments make a sum effect, a *condensed* narrative about decay or loss; in other words, the loss of control, the loss of the past, the loss of representation. I hesitate to call it fable, because it is so architectonic. Perhaps I should simply call it meta-fable, but the subject is how metamorphosis is built. The audience is supposed to sense the hand intruding.

So whatever one calls the process, the heart of its energy is the ani-morph, the twinkling monad—shape-shifting across dimension. Or even imploding—many substances into each other. The ani-morph is solid and absent at the same time. It is like a scar that narrates, a Braille of absences. The viewer can practically run a finger across the ani-morph's ridge, a very haptic sensation. The drawings leave an elegant wound as they dissolve to make way for motion.[9]

A shiver of lines escapes from a cartoon character's body. An earlier stage in production invades. We see the original drawing as a traced memory. Traces like this can be very unstable, like white noise invading. Any animated film has dozens of traced images that become ani-morphs. However, they are not merely slipups. They are pocket fables unto themselves, brief and easy to insert. What's more, when they dominate—as ruptures that fall into sequence—what results is a journey into an underworld. By underworld I mean a hidden place, an antiworld, where many atmospheres meet. Metamorphosis, then, is the surfacing of the hidden: entropy as molting, melting, melding, mutation.

Of course, this haunted and self-reflexive use of ani-morphing was never admired universally. Disney, for example, distrusted metamorphosis if it made the animator's drawing too obvious. A revealed scribble weakened the impact of full animation. In the words of Frank Thomas and Ollie Johnson, who have become the Boswells of Disney production methods, "When the animator distorts the figure, he must always come back to the original shape."[10] Donald or Goofy can be made to bulge and implode but must never lose their "personality," never turn into other *things* in the way Warner's characters did.[11] In cartoons of the thirties and forties, for example, there are no Disney gags where characters who slam into a wall turn into metal coins and twirl noisily as they land. That trick, so easily laden with ani-morphic frames, was reserved for Tom and Jerry cartoons, at MGM.

According to the Disney rule, once a character's body was shown—rubbery, watery, humanlike—its substance was irreducible. Walt was convinced that revealing the drawing behind the flesh could wreck the atmospheric effects that he prized so highly. He preferred wind, water, or heat to test the character's endurance. Disney nature made war with the character's body. In *The Band Concert* (1935), Mickey stays intact (no metamorphosis of any kind)—and on the beat—while conducting an orchestra thrown asunder by a tornado. His failure to morph was the central gag to the cartoon.

Pluto was perhaps the only Disney character allowed to show his scribbles—to have ani-morphs. For example, in a cycle drawn for the cartoon *Alpine Climbers* (1936), Pluto's body literally takes wing.[12] Lines snarl up until he looks like a bird in a blender, becomes briefly an ani-morph. His body appears to dissolve; that is, we see it lost for two drawings out of sixteen. However, the commentary on this drawn cycle by Thomas and Johnson advises us to turn away from ani-morphic lapses: "Never lose the personality of the character in either a long shot or a wild action."[13] Other kinds of ani-morph were treated in much the same way. At Disney, animators were told to avoid speed lines and rubber-band effects common to thirties cartoons—and used frequently by the Fleischers,

Tex Avery, and Bob Clampett. Disney was emphatic: clean up by shading; keep volumes constant.

Not that Disney did not want his cartoon characters to show off their plasticity; quite the contrary. But the tricks he thought would please the audience made ani-morphs nearly impossible. For example, characters were supposed to trip broadly, but slowly and gracefully. Goofy in particular often loses his balance so slowly that he seems to be moving in a tai chi exercise. He surfs gravity while he is lifted, then plunges two hundred feet. No matter how awkward the stretch, his body mass remains amazingly constant. His legs knot up like fishing line but never lose their mass—never a loose line to remind us of a flat drawing.[14]

By contrast, the Fleischer Studios in the early thirties (1931–1933) specialized in ani-morphing, with a simultaneity of effects that is still extraordinary to catalog; certainly by Disney standards, those effects seemed to wipe out the coherence of "story." Unlike Disney, the Fleischer animators liked to emphasize "traced memories" when they copied from live movement through rotoscoping (tracing live action into animation frame by frame). They also used allusion in a more self-reflexive way than at Disney—in other words, sight gags about media other than animation; for example, details drawn from vaudeville theaters and Coney Island rides the Fleischers knew. Or even traced memories of New York streets: kosher butchering in a bullfight scene; the Manhattan subway down a rabbit hole.

I suppose much of the difference came from the Fleischer's love of technology. The tricks that for Disney revealed the "illusion of life" (a caricatural naturalism) were for the Fleischers scientific marvels on display. That meant less commitment to hiding how animation was done. For example, in the Fleischer version of tabletop model animation, miniatures of caves or streets are visually unmistakable; they are much rounder than the cels placed in foreground. What Fleischer called "3-D" looks much more constructivist than Disney's multiplane effects a few years later. In multiplane, the cels are placed in slots that make them look like cutouts in deep focus, but they are smooth spaces from front to back. In Fleischer, however, the glass window of the diorama is plain to see; it softens the 3-D background severely, leaving the flat drawings—in front of the glass—very crisp and ripe.[15]

Betty Boop's Snow White (1933) is undoubtedly Fleischer's animorphic masterpiece, particularly its final sequence in an underworld—it is both an Orphean journey (as in the myth of Orpheus) and an orphic journey (a silly dance of death set to music). Inside this underworld, animorphing governs movement and motivation. For example, the evil queen turns Koko the Clown into a shape-shifting ghost while her mirror keeps sprouting hands, and she conjures a blackface to tell her who is fairest of them all. At the same time, Koko as ghost is rotoscoped from a clip of Cab Calloway. Of all the Fleischer characters, Koko was rotoscoped the most often. By 1933, it gave him a phantom presence, too often invaded. Graphically, rotoscoping leaves scars—something a bit too human, a bit too lithe, subtle but plain to see. Koko practically inhabited two bodies at once, from a cartoon clown who shuffled (buttery head, sacklike body) to a leaner man who ran gracefully (more angles to his chin; a stiffer spinal column). Koko was designed to be haunted, wrapped in billowy cloth that was ideal for a ghost dancing between bodies, particularly in this, his last extended appearance, his swan song.

Koko sings "Saint James Infirmary" while turning into a twenty-dollar gold piece, then into a "shot of that booze."[16] At the same time, to illustrate the line "crap shootin' pallbearers," the wall behind him is lined with murals of skulls and cows together, gambling. That bears scrutiny, usually requires a few viewings: it is intentionally *traced* like the wall of a Coney Island Mystery Cave Ride. It is also traced out of a collective imaginary (at least the collective of animators). The skulls of African Americans reenact the greasy underworld of back-alley and saloon life in Harlem. But not Harlem as blacks knew it—this is Harlem as the white male Fleischer animators sensed it. The skulls resemble the racist extremes in Currier and Ives prints, with Jim Crow white-on-black "pickaninny" scowls, and the ooga-booga lips common to American cartoons until the late forties.

The scene is rich enough in allusions to New York—as the animators lived it—to suggest a trace memory, like a foldout postcard filled with racy sketches of scenes in the city, a composite of weekend leisure for the boys at the Fleischer Studios. It is their boozy Manhattan caricatured in some detail, as an inside joke. (On Fridays, the animators used to visit hot spots together, particularly Earl Carroll's Vanities, the Ziegfield Follies, wrestling, and Hoochie Koochie dance clubs—and of course the Cotton Club.)

Even Betty's body was a traced composite—a traced memory—of women they saw along the way. Her garter was like those favored by Hoochie Koochie dancers so popular at burlesque and dance parlors.[17] She slouched her back like a flapper at a speakeasy. Her banjo eyes and her bounce were copied from the moves of vaudeville singer Helen Kane. Her head bobbled like a Coney Island kewpie doll shaking on a spring.

The plot, such as it is—more a container than a plot—turns a Mystery Cave ride into a blend of Coney Island and Manhattan. It proceeds like a taxi tour, a few drinks at each stop. First Betty enters during an opening Ziegfield chorus number, until the evil queen orders, "Off with her head." (Another ani-morphing gag: the evil queen's thumb and forefinger turn into a guillotine.) Then, while tied to a tree, Betty torch-sings "Always in the Way," as if she were in a vaudeville "mellerdrammer." But very quickly, she breaks free. While walking downhill, she trips absentmindedly, rolls into a snowball, and slides into an icy lake. While frozen she keeps sliding, passes through the Seven Dwarves' cottage, and into a Coney Island ride. Or should I say an amusement park underworld/morphworld, even with a potted plant on her coffin, to remind us again of New York apartments, where windowsills were decorated with flowerpots.

Meanwhile the evil queen is ani-morphing her way to the cave. She runs her body through the mirror as if through a hoop and transforms into a hag witch, forces Koko to shape-shift, and freezes Bimbo the Dog. But even this witchcraft fails to kill Betty. In frustration, the evil queen turns into a cakewalking, rather cute dragon, with ducks on her head who honk like bird whistles for geese hunting.

The peculiar heat from her morphing into a dragon also melts the underworld, releasing Betty and her friends, as if they were mammoths thawed from the ice. A musical chase ensues, climaxing with an ani-morphic gag as spectacular as any the studio produced. Bimbo grabs the dragon/witch by the head and turns her inside out. Her skeleton is visible in black, as if she were wearing tights painted to look like a skeleton for the scary finale onstage. It is easy to run this skeleton gag frame-by-frame on video: her dragon body melts, then seems to run off by itself—pauses in ani-morph—while the skeleton makes a three-quarter turn. The way she turns resembles gimmicks in theaters—musical finales on a revolving stage. Anyway, that finishes off the queen. With a last downbeat, Betty, Koko, and Bimbo flee the cave, and do a May Dance. The ani-morphing underworld is gone, but the characters show no sign of wear, give no sense that this was any more than a theatrical journey, despite all its allegorical layers—which brings me to broader questions.

The through line, such as it is, has to be called metamorphic. After all, Betty and Koko's journey is crammed with versions of shape-shifting, body to ghost, frozen death to life, flesh inside out, a world outside caving in. The animation, even the ani-morphs, fit narratively into a very thin musical sketch based on a fairy tale. In fact, the ani-morphs (rotoscopes, skeletons, and ghosts half transformed) flesh out the absence of plot—a vaudeville tour through the underworld of New York entertainment: cardsharping, running craps on the street, speakeasies in back doors, boating rides under the sign of death in Coney Island. The "story" is about uncertainty—modernity and the depression as the Fleischer team witnessed it. Indeed, from 1931 to 1933, Fleischer cartoons have a peculiar bite to them.

Bimbo's Initiation (1931) is an earlier example. Bimbo, as if trapped in immigrant panic, is forced to spin into a labyrinth of imprisoning rooms. Some rooms sprout knives that try to stab him; others grow mouths that gulp him; and others erase gravity and force him to crawl across the ceiling. (The comparison with Kafka's Gregor Samsa seems unavoidable, although the Fleischers knew nothing of European modernist literature, or surrealist theater and film; theirs was a homegrown pathology of urban life.) Bimbo keeps refusing to be a "member" of what seems like a strange Bundist or Masonic order in caricature—hooded men with spent candles on their heads. This was also an era that had witnessed a huge revival of the Ku Klux Klan. Finally, one of the leaders pulls off his hood and turns out to be a lady poodle, the sexiest poodle Bimbo has ever seen, "a pip," he calls her. She does a bumptious bump and grind for him. He grins as if he were being tickled from the inside out, lasciviously; his eyes follow her: she is an earlier version of Betty Boop. Then all the hooded KKK Bundists take off their hoods. They're all copies of this voluptuous Betty Boop. Bimbo slaps her ass; she slaps his—a raunchy version of the dance black bottom.[18] A gleeful layer-cake chorus-line finale ends the short in the way most of the Fleischer cartoons of that era ended: like a Victor recording that runs out of threads and simply stops, on a final trumpet or downbeat.

This is a dark piece of work for kiddies to watch, even for adults. And there were others almost as dark from Fleischer in the twenties, before the depression. In two of the most remembered, Max gets multiplied industrially and attacks Koko in *Cartoon Factory* (1925), and the world explodes and New York goes cockeyed in *Koko's Earth Control* (1928). The Fleischer's apocalyptic cartoons can be periodized, from milder to hot, like a salsa bar. By the mid-thirties, under censorship during the Shirley Temple era, their macabre twists continue but are more about repression, more about guilt than dancing on your grave. Two cartoons in particular from 1936, *Cobweb Hotel* and *Small Fry*, have since become cult favorites in Weird

Cartoon collections. They each are built around nightmare chases like the Boop cartoons—many metamorphic scenes—but here instead of bouncy dance numbers, we see flies tortured, or baby fish forced to swim through inky inversions as if they were facing Lethean death. Although Fleischer employs fewer morphing gags throughout the thirties, and tries to make cartoons that resemble Disney full animation much more, something of the allegory of the ani-morphic underworld remains.

I have always assumed that the Fleischers' insistent diablerie came out of the immigrant world they knew from Bedford-Stuyvesant, Brooklyn, as children. They understood the xenophobia of a Jewish neighborhood, living as part of the largest influx of foreign immigrants to hit any American city at that time. In New York, as in Los Angeles today, up to 25 percent of the total population were foreign born. This, in itself, is metamorphosis—ethnic cultures caught "between," in ani-morphic shock. This agonizing unease is heightened by economic disasters like the Great Depression, or by class warfare, strikes, and street fighting. But let us put aside the grand overview and imagine instead the look and feel of streets on the way to work; the muddle of urban legends about gangsters, cops, anti-Semitism, and racism; even the rumble of the elevated train; the childhood friends who do not cross over into middle-class jobs; the taste and service of the food. Consider the way a character is drawn, to match this mood of instability and paranoia, how the body has "attitudes" that express how heavy or light the clothing is, how men and women use their arms as they walk through a New York street. The subtleties of urban paranoia have their own visuality; a rhythm in the shoulders, in the posture, in walk cycles; angling the neck just a fraction to watch out for who may be behind you; staying toward the outside of the sidewalk. It is not really a fear of crime so much as a fear of the mixing of classes and races. It is a comfort zone built out of a mood of uncertainty.

This streetwise paranoia is very evident in the mordant edge of Fleischer cartoons, and it is apparent in a different sense today. I am reminded, of course, of the shock waves in Los Angeles from 1992 through 1994 (looting, fires, earthquakes, massive recession). Again, let us put aside the broad picture of 1992 for a moment and consider how the economic aftershocks spread into less obvious spaces: panicked suburban home sales; into how streets are patrolled, how people shop at night, how the politics of class and race are erased by the daily routine. The sheer banality of controlling one's fear is the world drawn by the ani-morph. Imagine how someone tiptoes briskly from their car after parking at night in an empty lot. The walk is not an expression of "paranoia," merely how one moves to establish a comfort zone, when the space feels "ani-morphic." Then imagine how the movie industry turns these anxious (ani-morphic?) zones into cash, into scripts where computergraphic monsters, as evil immigrants, destroy middle-class real estate. Obviously, I am thinking of nineties f/x blockbusters such as *Independence Day* (1996), part of a bumper crop of disaster films inspired by the shocks of 1992 to 1994. Like Fleischer cartoons, these are ani-morphic fables. That is, both Fleischer shorts and bloated disaster pics are allegories about powerlessness, about alien presence, about underworlds where the animator builds social imaginaries about paranoia. The sources can be a bit laughable, I admit. I imagine the power brokers in the film industry from 1993 to 1994 watching houses burn in Malibu, or their pool spilling over while their best china explodes during the earthquake. Then, after the shaking subsides, or the fires are put out, they head for a local watering hole in West Hollywood, perhaps, and decide to green-light any film that sounds like a special-effects disaster epic. Somehow, lava feels right, anything that shakes like a vengeful mountain god.

But before launching into the present, and some comment on the current politics of morphing, I have to do justice to the ani-morph. Its range is staggering. For example, during the seventies, even before post–*Star Wars* giganticism, there were dozens of ani-morphing shorts. I will cite two, to stand in for many.

Jan Svankmejer's *Dimensions of Dialogue* (1971) is encyclopedic in itself. An Archimbaldo creature (a homunculus made of fruit) eats a hominid made of industrial parts. Through mad pixilation (speeding up live action), they chop and dice each other ever finer, until they turn into clay sculpture—which segues into the next scene, where clay lovers start off by making love, merging their bodies into a heaving ani-morph. Then they take a break, reform into their bodies, and smile. However, an ani-morph is left behind. A little blob of clay tries to find a home in their bodies, but they kick it away angrily. Finally, the blob becomes the nuisance that sparks a battle to the death between them. The once loving couple literally gouge their clay bodies into a lumpen gravy. This brown heap acts as a cross dissolve to the third act: Two beefy clay bureaucrats try to communicate diplomatically. Out of their mouths, instead of words, they present shoes to be tied, bread to be buttered, toothbrushes to be filled, and a pencil sharpener gripped neatly by a highly salivated tongue. However, very soon, they lose track of which object is which. They start buttering the shoes, pencil sharpening everything. At last, hyperventilated, they collapse into doughy lard cakes, like melted bulldogs.

A quick review of techniques: Svankmejer mixes up his ani-morphs, using substances from clay to everyday objects; that is, by medium, nimbly leaping from food to thimbles to clay to toothpaste tubes. Each substance becomes a robotic piece in a theater of war—an ani-morph to mark the precise instant when social codes dissolve into mindless traces. Then the insanity escalates into oblivion. The key to his condensed fable is the mix of the crafted, the industrial version weirdly stuck inside the organic version, and the melee that follows: Archimbaldo vegetables, the chopper/blender effect, clay diplomats with rhino necks and watery green doll's eyes.

Another film rich in ani-morphic vocabulary is Caroline Leaf's *The Street* (1977). Done in fingerpaint on glass, it animates a story set in the Montreal Jewish ghetto during the 1950s. A small boy waits for his grandmother to die so that he can get her room and not have to share with his sister. But his grandmother stubbornly hangs on for months. Meanwhile neighbors and relatives make sympathy calls, carrying with them a montage of textures: the dense outlines of buildings, the layers of flesh and clothing in a crowded apartment. Leaf often uses the ani-morphs as cross dissolves: from neighbors' bodies to buildings, or as fly-throughs before the age of the computer, dissolving paint in the path of children running down the street. The neighborhood is soaked in earthy browns and jaundiced yellows, both nostalgic and suffocating. The boy is trapped inside his vigil, then frightened when his grandmother finally dies. To remind him of his greed, the walls of her room are still textured by her presence. The space is infected by memory, as in ani-morphic films by the brothers Quay, for example. Finally, the bodies of mourners swarm into the house on the day of the grandmother's funeral. Their noise transfers for the boy into vaguely competing visual textures, fingerprints of memory crossing each other against the sheer angles of the house.

Here too the ani-morphic tension between the organic and the industrial object makes for a bizarre statement, like the Brundlefly in Cronenberg's remake of *The Fly* (1986). Finally, poor Seth Brundle becomes a brooding mutation of metal and tusk, with barely a trace of the human left, or even of the fly. He has been mutated into a hybrid that defies both the industrial and the organic, yet is frighteningly compelling, because we sense all those phantom human and fly limbs: the traces of genetic activity; bits of conscious will diffused into uselessness, limping, groaning, and collapsing of its own weight.

Ani-morphing becomes an allegory about the organic disappearing into the industrial, as in the Quays' *Street of Crocodiles* (1986). That masterpiece deserves an extended note here. The Quays very consciously employ a kind of ani-morph. In 1995 they said: "What's good about film is that while you're moving through space, it opens up these little parentheses and the imagination drifts off and is flooded by these contaminations . . . we love that vague wandering off."[19]

Similarly, the credits at the end of the film are a clue. The film is excerpted from the novel by Bruno Schulz, *The Street of Crocodiles* (1932)—a source even richer in ani-morphic imagery than the film itself. The novel is about a rotting Polish town haunted by the decline of a gold rush years before, among other disasters, particularly at the severely blighted Street of Crocodiles. This becomes the allegorical setting for the film: a street so embarrassing to the other townsfolk that its precise location is omitted from the map of the city. There is only a white blank where the street should appear. Even the mapmaker had been afraid to visit the street. It features cheap jerry-built houses everywhere, suitable now only for the scum of the city. The streetcars are made only of papier-mâché. The coaches run blindly without drivers. The only "concession to modernity" are collages cut from yellowing old newspapers.[20]

The way collage is applied by both Schulz and the Quays is very useful here—it takes us on an ani-morphic journey through an underworld of contrasting atmospheres. Let me explain with one example among many, a memorable image in the film: butchered organs simply manifest inside a dusty industrial ruin. Kidneys and then a liver morpho-magically appear in haptic collage during a ritual conducted by dolls with Holocaust-like numbers imprinted on hollow, empty skulls. It is a nonfertility rite—ani-morphic, as in a ritual where both organic and inorganic are lost memories.

Whenever I teach this particular film, I always read as well from Schulz's ani-morphic descriptions of supernatural plant life in the village. On another street, an alien algae grows in thick colloids that can be preserved with kitchen salt. However, the cell structure of this plant remains utterly "amorphous," without cell walls or nuclei. As a result, all this grieving slime can do is imitate what lies nearby (another ani-morphic act). Elsewhere, a strange mulch attacks furniture, until table legs start to grow like trees. These amnesiac or aminoidal tables try to relearn photosynthesis, much the way the streets try to remember who lived there, through blind rituals. In another organic/inorganic blind ritual, the branches of the furniture climb the walls like a tree and strain against the limbs of the house. The groans suggest a building about to have a heart attack. The furniture misremembers, much the way the dolls do. The joists of the house are crippled, the impulse out of touch with the intention.

In the world governed by ani-morphs, brute matter comes to life but has forgotten why. Instead, it languishes somewhere between discovery and annihilation, fails to connect. The underworld of ani-morphs grows into an ambient space, an epic narrative without a memory. This technique of ani-morphic pauses resembles Deleuze's description of René Clair's use of Vertov, for the film *Crazy Ray:*

> The desert town, the town absent from itself, will always haunt the cinema, as though possessing a secret. The secret is yet another meaning of the notion of interval: it now designates the point at which movement stops and, in stopping, gains the power to go into reverse, accelerate, slow down.[21]

For the Quays, ani-morphic intervals are enacted, rather than acted out, by dolls and a puppet. Each performs without memory, but with precision. The dolls' heads sew. The puppet (with a face like the great installation artist Joseph Cornell) shares in their ceremonies. They all act as if awaiting orders that never arrive. They pause for cues, tilt their heads, then burst into action anyway. Perhaps they are in the shadow of someone else's long-gone story? Their world is out of scale. The dolls are miniaturized inside an unknown maker's creation. A massively oversized pair of scissors has been left on the glass roof. The sewing factory has rusted. No humans seem to work there anymore. The cordage machines spin like ancient film projectors. Occasionally, die-hard spectators show up on the corners of the screen (as in all Quay films). But the voyeur seems no more certain of direction than the others; he too is a puppet—a very important distinction for stop motion animation. Puppets are another species of "character" than dolls. The puppet's face is weathered, historicized: thoughtful, greedy, scheming. The doll, on the other hand, is merely a remnant of a childhood desire; it is a carrier, much blanker in expression, and clumsier in its expressive movement.

The poet Rilke, another unlikely source, may be useful in clarifying the distinction. In one of his odd ruminative essays, Rilke remembers the scrawny expressionist wax dolls made by Lotte Pritzel from a show in

1913. "The doll was so utterly devoid of imagination that what we imagined for it was inexhaustible. . . . The doll was the first to make us aware of that silence larger than life which later breathed on us again and again out of space whenever we came at any point to the border of our existence."[22]

In stop motion animation, puppets exaggerate the presence of animorphs. As early as 1912 in Starevich's films, which essentially begin the trajectory that leads to the Quays' work, each body that is animated leads you closer or farther from memory—from insects to onion skins to *pulchinelle* dolls. We observe their source and watch them blindly follow, through pauses where they seem to be listening, as if in a reaction shot. They occupy two places at once: the space of production (the hand of the maker), and the space on film. But in the Quays' film, these spaces seem to refuse narrative revelation. Their sum is a stop motion "parentheses." It acts as if "contaminated" by a lost narrative. Movement proceeds very slowly and precipitously, the way water leaves a leaky jug—but always in mid-morph: again, the anatomical scenes using cut organs from the butcher; scenes where heads are transferred from one body to another, from doll to puppet.

An extraordinarily rich interior allegory is inscribed here. In *The Street of Crocodiles* layers of ani-morph jump from phantom memory to phantom film production, a journey through an underworld inspired by Schulz's Kafkaesque magic realism, by Svankmejer's surrealist animorphisms, by the rusting of industrial sites (from Pennsylvania, where the Quays grew up, to London, where they work). Finally, the uneasy memory of the lost Jews of Poland seems represented here as well; this is a diorama "contaminated" by the Holocaust. And yet the structure of the scenes is astonishingly coherent. Place, character, and plot coexist as animorphs between a past that is forgotten and a future that is a memory trace.

Most surprisingly, this film (and the Quays' work generally) has been embraced by the MTV generation. The Quays have produced a number of commercials, even some for MTV. A video for the band Tool clearly plagiarized the Quay technique, almost as homage. I also find that student computer animators adore the Quays' work. When I ask why, the students reply essentially that they want their films to be haunted also, as an antidote to the hygienic digital screen.

The Quays—and the ani-morph—can be seen as an alternative to the disembodied sensibility that the computer represents, and to the poetics of loss that it should reveal but rarely does as yet. But I would rather not oversimplify the matter. We need prophetic allegories about the global corporate civilization, also a fiercer poetics about identity dissolving, beyond the usual debates about public and private space or about the posthuman body. One way to proceed, among many, might be to produce a cinema of ani-morphs that reveals production methods far more.

By contrast, on the computer, how might the ani-morph work? I listen to a brief debate between Christine Panushka, an animator who specializes in metamorphosis, and Vibeke Sorenson, one of the pioneers in "morphing" software, also a computer animator. Sorenson insists that *interpolation* on the computer (the elegant cross-dissolving of shapes) works the same as do drawn ani-morphs. Panushka is not as certain. Indeed, many in the animation industry wonder if computer 3-D morphing programs are a mixed blessing; 3-D applications tend to smooth out glitches and hesitations more than 2-D ani-morphing. Their effects display less of the mercurial, the lightning hand, traced memories, allegories of entropy and ruin. Furthermore, it is clear that the workplace in globalized image industries tends to abhor the ani-morph as dogmatically as did Disney. Although results can be very elegant, their clean looks are so cybernetically cute, so close to blasts of ad copy, that they become more an allegory about impulse buying than ani-morphic hermeneutics.

I surf the Web for an hour, to see what the cyber-mall has to say about contemporary metamorphosis. Dozens of morphing programs are for sale, from about $900 to as low as $20, most of them extensions of "feature-based image metamorphosis" software developed in the early nineties.[23] Morphing has escaped from its animation ghetto and now is vital—not only for mainstream narrative cinema but also for banking, architectural planning, engineering, advertising, and interface design. In the De-Evolution Gallery, your family portraits can be morphed back to Neanderthal. The book *Animorphs* by K. A. Applegate gives readers a sense of what it would be like to be another species.[24] The name *Morph* (originally the name of a character on BBC from 1981 to 1983) has taken on a pioneering meaning, as in "Morph's Outpost on the Digital Frontier."

The technology has now outraced the vocabulary. Contemporary morphing looks increasingly like an allegory about accommodation, an attempt to turn our 2-D sense of decay into 3-D global fantasies. Perhaps this is a "natural" evolution in that, one could argue, computergraphic morphing is always an epic form of evasion. It is an empty display of worlds in collision, of new species of identity that are perhaps no newer than the assembly line. But most of all, it pretends that we can author our own modernity, not only survive the shocks but run ahead of them. The more I consider what morphing "says," the more conservative its message begins to sound to me.

That is why I make such a fuss about this term *ani-morph.* It reveals a crack in the armor of metamorphic plenitude, the point where the myths of outracing modernity fall apart, and we see the crisis of identity more clearly. The ani-morph can be something of an antidote. I would hate to see it lose its edge. It may be to our computer culture what the dissolve was to film culture—simultaneously an erasure and a transition.

Notes

1. These are also called "quick sketches." Film historian Donald Crafton initiated much of the academic study of lightning hand. He expanded his articles on the subject in *Before Mickey: The Animated Film, 1898–1928* (Cambridge: MIT Press, 1982). The term "lightning drawing" has become fairly standard in critiques of student work at various film schools (at USC and Cal-Arts, certainly). In other words, the problems suggested by linear metamorphosis remain fundamental to the field, even today, with the daunting presence of the computer.

2. *Humorous Phases of Funny Faces,* J. Stuart Blackton (1906).

3. *Little Nemo,* Windsor McCay (1911). See: John Canemaker, *Winsor McCay: His Life and Art* (New York: Abbeville Press, 1987), 132. *Little Nemo* was produced by Vitascope, Blackton's company.

4. Interview with Ward Kimble, July 1987.

5. Chalk remains a useful metaphor here, at least as an excuse to play with the material further, add a few terms: Chalk can be erased, broken into dust, shaded by hand. It has texture, facture, sound, what can be called the haptic (tactile, synesthetic). The haptic is essential for all ani-morphed line, for all special effects, in one of two categories: it looks either anabolic (turning food into tissue) or metasomatic (rocks changing substance).

6. I realize that *ani-morph* is only a letter away from *anamorph,* and *anamorphosis.* In future essays, I will explain how close and far away that is.

7. I should add that this metamorphing background is fundamental to the gag structure in cartoons, what I call the "controller" element; that is, Bugs or Droopy always get there first, but Daffy or Wolfie suffer in the wrong atmosphere, where existential anomie governs.

8. Walter Benjamin, "The Origin of German Tragic Drama," trans. J. Osborne (London: New Left Books, 1977), 178–79. These are the classic pages, so often

cited, discussed in detail by Susan Buck-Morss, by art critics Benjamin Buchloh and Craig Owens.

9. This is similar to what I call "distraction" in *The History of Forgetting* (London: Verso, 1997).

10. Frank Thomas and Ollie Johnson, *Disney Animation: The Illusion of Life* (New York: Abbeville Press, 1984), 138.

11. Personality was a very specific term for Walt Disney, the point where a character—with personality traits established by the story department—went through a cartoon conflict, reflected in the graphic design, rhythm, and colors. But personality also had a second meaning during the thirties (and afterward, on TV, for example, often called a "personality" medium). This was less discussed at Disney, simply understood—that Mickey or Donald were star personalities to license outside of films, in toys, watches, and finally in theme parks.

12. Thomas and Johnson, 148–49.

13. Ibid., 149.

14. See *Moving Day* (1936) and *Clock Cleaners* (1937), among the best of Art Babbitt's renditions of Goofy, as well as the Sport Goofy series directed by Jack Hanna in the forties and fifties.

15. The most widely noticed examples are in *Popeye the Sailor Meets Sinbad the Sailor* (1936) and *Sinbad the Sailor* (1939). Many other shorts by Fleischer in the thirties used tabletop miniatures ("3-D Process"), including a Boop in color, and the features *Gulliver's Travels* and *Mr. Bug Goes to Town.* In all of these, the 3-D is only in a few scenes.

16. Ani-morphs from this sequence were isolated in *Seven Minutes* (London: Verso, 1993); see, for example, 79, 93.

17. *Hoochie Koochie:* a pseudo-Egyptian belly dance that was popular at burlesques and "Hoochie Koochie" parlors.

18. On "black" bottom: The dog and mouse characters of that era, including Freleng's Bosko for Warners, the early Mickey Mouse, and the Fleischers' Bimbo, often show traits that suggest black men. These mannerisms are mixed, of course, with those of white males (the voice, the plots). It is another peculiar coding of black to white, here as a trope where domestic animals mutate almost into humans, but never entirely. It also identifies the deep presence of black dance, music, and theater in the sources for these cartoons.

19. Carolyn Steel, "Space That Breathes," *Blueprint* (October 1995): 42.

20. Bruno Schulz, *The Street of Crocodiles,* trans. C. Wieniewska (1934; reprint, New York: Viking Penguin, 1977). The Quays used a different translation, clearly, since the quote at the end of the film does not match the same passage here (see p. 110).

21. Gilles Deleuze, *Cinema 1: The Movement-Image,* trans. Hugh Tomlinson and Barbara Habberjam (1983; reprint, Minneapolis: University of Minnesota Press, 1986), 83.

22. Rainer Maria Rilke, "Dolls: On the Wax Dolls of Lotte Pritzel," in *Essays on Dolls* (London: Penguin Books, 1994), 32–33.

23. See Thaddeus Beier and Shawn Neely, "Feature-Based Image Metamorphosis," *Computer Graphics* 26, no. 2 (July 1992): 35–42.

24. K. A. Applegate, *The Invasion,* Animorphs, no.1 (Bergerfield, N.J.: Scholastic Books, 1996). This is the first of twenty popular children's books about a species known as "Animorphs."

Figure 3.1 Three Olmec figures representing the transformation from human to jaguar. Heights, *left to right:* 19 cm, 18.8 cm, and 8 cm. Courtesy of Dumbarton Oaks Research Library and Collections, Washington, D.C.

3

Morphing **Magical Transformations** and Metamorphosis in Two Cultures

LOUISE KRASNIEWICZ

Picturing the Magic

In 1996 the magical transformation of an ancient shaman into his jaguar alter ego was dramatically depicted by a series of small figurines in a National Gallery exhibit of pre-Hispanic Olmec art.[1] Juxtaposed for the first time, three thousand or so years after their production,[2] these small statues from one of the earliest Mesoamerican cultures were notable for their illustration of the entire process in which a human transforms into an animal as part of a shamanic journey to a sacred place (see figure 2 for statues 3, 5, and 6 in the sequence). So dynamic was the sequence that it suggested to one scholar "illustrations from an old-fashioned flip book or, better still, individual frames from a film clip of the transformation sequence in a 1930s were-wolf film."[3]

The first three statues in the sequence have the same posture, fisted hands on knees, sitting back on the heels but leaning forward. They show the beginning of the transformation in which a human begins to acquire feline characteristics. The first figure is clearly a man, bearded and with long hair, but no other marker of transformation. The second figure shows signs of change. Now the head has a carving of the toad that is the source of hallucinogenic drugs used by shamans in rituals of transformation. It acts, says F. Kent Reilly, as a "symbolic verb," telling us the activity (a transformation) that is taking place.[4] A design on the bald head indicates that the skin is beginning to split open, revealing a creature underneath.

The third figure even more distinctly shows a transformation under way. Still seated like its more human predecessors, this creature has feline features except for its torso, legs, and ears. It now has a snout, fangs, feline ears, a molting tear on its shoulder. The fourth figure retains a human body but is now rising from its seated position. Its head is fully feline. Finally, the fifth and sixth figures are carved standing with the last one completely transformed into a were-jaguar, a cat in an upright position, complete with tail, but its hands and stance human.

A few years before the Olmec transformation sequence was displayed in a small case in the nation's capital, television viewers everywhere witnessed a similar magical transformation. In what has become one of the best-known early examples of computer-generated "morphing," in 1991 the computer graphics studio Pacific Data Images (PDI) designed an image sequence in which a swiftly moving automobile transformed itself into a gracefully running tiger. In a sequence that lasted only a few seconds but changed the course of television advertising, the car zoomed up a steep mountain road, and then its body rippled, developed stripes and legs, and finally merged with that of the tiger, which continued the motion across the screen. Creating this advertisement for Exxon gasoline required PDI to convincingly morph two objects that shared no common physical features or logical connection, an idea they originally judged as "a bit weird."[5] The challenge for these digital wizards was to convince the viewer that the car had actually changed into the cat rather than just replaced it on the screen. The tiger was no longer just in the tank, and the "morph" became a stock effect, a visual metaphor that could connect any two unrelated elements and combine them for mass consumption.

These two stories of the magical transformation of nonfeline entities into magnificent predatory cats, although centuries apart in their executions, speak to the similar fascination their respective cultures have with shape-shifting. Despite their differences in cosmologies and technologies, these cultures seem to be relying on similar stories to address concerns that span centuries and traditions. Humans have long marveled at, revered, feared, and suspected those figures who, in our stories and art, have been able to accomplish the unsettling change from one form of being to another. From the Greek myth of Circe, who turns the soldiers of Odysseus into swine, to the Celtic tale of the Children of Tuireann, who change themselves into hawks, to the ancient Aztec story of Quetzalcoatl, in which rival deity Texcatlipoca transforms himself into both an old man and an ocelot, to the Polynesian Queen Kawelu, who becomes a butterfly and dies, heroic and larger-than-life metamorphosis abounds in the preindustrial non-Western world. These tales of metamorphosis are about power struggles, initiations, creation and rebirth, sex, love, and war; in each case, transformation provides a way to connect separate worlds or states of being, to cross boundaries of space, time, and existence. Indeed, the depiction of humans who change into animals, animals who shapeshift into each other, or inanimate objects that morph into live ones are as old as human culture itself; some of the earliest human art, in the form of cave paintings that are thirty thousand years old, show man-beast combinations that suggest a metamorphosis in progress.[6]

The widespread existence of stories of shape-shifting and magical transformation points to the likelihood that they address some of the fundamentals of human social life, perhaps questions about defining humans and their struggles to remain human. In this continuum over time and space, humans have used the concept of magical transformation to speak about their most dire concerns and compelling interests. The myths and metaphors of metamorphosis and transformation have been useful in other cultures for making sense of their circumstances, for giving form to their beliefs and activities, and for communicating these concerns to each other and to the inhabitants of other cosmological planes. In a culture like the Olmec in ancient Mesoamerica, metamorphosis appears to be a core concept that shaped not only the Olmec civilization but the entire Mesoamerican complex that developed until the sixteenth-century incursions of the Spanish.

Does contemporary computer morphing follow in this tradition? Historically, metamorphosis has been associated with low-technology forms of communication—myths, fables, and rituals, with the spoken word and its artistic manifestations. Today, morphing visuals are created with the most sophisticated computer technology and presented through the myths, fables, and rituals associated with our cinema, television, and computers. Although the differences in technologies should be no barrier to a cross-cultural understanding, the vast differences in cosmologies suggest that contemporary representations of computer morphing are not simply reiterations of traditional tales of metamorphosis.

As we shall see, the Olmec, through their sophisticated and fluid art forms that emphasized metamorphosis, used "the power of images to embody, codify, and communicate shamanic ideology, ritual systems, and political charter."[7] We, on the other hand, promote a limited view of magical transformations in our reliance on computer morphing; we use it as an aesthetic, seemingly apolitical form of expression and as a way to demonstrate our proficiency in, and mastery of, advanced computer techniques. Yet morphing has the potential to express the same wide-ranging concerns that tales and images of metamorphosis do in other cultures. By comparing and contrasting these two traditions, and by looking at how others have incorporated concepts of transformation into the very structure of their being, we can gain some insight into the possibilities that our new, high-tech versions of this ancient visual form have to offer our culture. In addition, because we have learned that the myths and metaphors of transformation were useful to the people of other cultures for making sense of their circumstances, a look at our fascination with digital morphs can tell us something about ourselves and our own circumstances at the turn of the millennium.

The Shaman's Journey

In other cultural contexts, metamorphosis, the magical change of physical form, is a powerful mystery, "an experience outside the realm of real life that has nonetheless persistently captured the human imagination."[8] In this mythic view of the universe, metamorphosis is dependent on an acceptance of the possibility that the divine can be embedded in the real, that the different realms of existence, however they are locally defined, can cross over into each other.[9] Real and unreal, phenomena and essence, this world and the other are mixed. The images of metamorphosis that other cultures promote are therefore portrayals of that moment when the gods enter and consort with the natural and common, and humans exhibit the power to explore and influence both the natural and unnatural worlds. This hints at a richer, more complicated world than the one we walk in daily, a world whose passage requires illusion, imagination, and fantasy or at least the willingness to consider imagery that "sets the viewer's psyche in motion, reveals arbitrarily rather than describes thoroughly, disturbs more than it satisfies, and strongly suggests the impossibility of seeing everything at once."[10] Images of metamorphosis are designed to do just that.

The stories, myths, images, and experiences of metamorphosis are often associated in non-Western cultures with what have come to be called "shamanic" activities. The term "shaman" originated in studies of hunting and herding Siberian groups to refer to their religious specialists, but in anthropological research, it has come more generally to refer to those figures who provide contact with spirits of other worlds through soul journeys that promote healing, revenge, acquisition, and other practical activities.[11] Shamanism is "an ancient practice of utilizing altered states of consciousness to contact the gods and the spirits," a visionary tradition in which the journey to sacred places yields information of cosmic importance.[12] The shaman is himself or herself a figure of metamorphosis. As Nevill Drury explains in his popular description of shamanism:

> We can define a shaman as a person who is able to perceive this world of souls, spirits and gods, and who, in a state of ecstatic trance, is able to travel among them, gaining special knowledge of that supernatural realm. He or she is ever alert to the intrinsic perils of human existence, of the magical forces that lie waiting to trap the unwary, or which give rise to disease, famine or misfortune. But the shaman also takes the role of an active intermediary—a negotiator in both directions. As American anthropologist Joan Halifax points out: "Only the shaman is able to behave as both a god and a human. The shaman then is an interspecies being."[13]

Although shamanism as such does not exist as a unified concept or religion, shamanic ideas borrowed from non-Western cultures have gained tremendous popularity in Western cultures in recent years. Neo-pagan and New Age religions promote shamanism (which they often claim can be "learned" in a weekend retreat) as a way to achieve personal satisfaction and knowledge, and as a way to heal the self and the world. But the traditional shaman does not have a personal agenda for his or her dangerous and often horrifying excursions into other realms. From the moment he or she is called to, or chosen for, this profession, the shaman becomes a servant of the culture. The shaman's job is to bridge the gulf between different states of being, to act, as Peter Furst suggests, "as a mediator between the visible and unseen worlds."[14] Journeys to the other layers of the world are necessary for a society's well-being and continued existence. As the "specialist in crossing this otherwise impassable gulf,"[15] the shaman also "expresses the possibility of coming together again,"[16] of renewing the culture. Shamanic activities exist in different religious contexts, but "shamanic logic" involves concepts of a layered cosmos in which we occupy merely one layer. The other layers are occupied by spirits–of ancestors, deities, or nature–who do not share our form of being even as they share our space and time. These spirits affect our world and for this reason must be contacted, and then controlled or placated, by the shaman.

Computer morphing entered the toolbox of digital artists in the 1980s when the visual effects company Industrial Light and Magic used the technique in the movie *Willow* (1988) to change a sorceress into a series of animals.[17] Willow, a somewhat inexperienced hero, waved his magic wand over an enchanted opossum, mistakenly changing it into a goat, an ostrich, a turtle, and a tiger before successfully returning his ally, the sorceress Fin Raziel, to her human shape. That the technique was introduced to the public in a film with mythic and fantastic themes was an appropriate birth for a new imaging technology that, like mythical metamorphosis, simulates the unsettling change of one form of being into another. This "shape changing metamorphosis scene . . . caused an immediate sensation"[18] and led to the widespread use of morphing scenes in subsequent movies such as *The Abyss* (1989), *Terminator 2: Judgment Day* (1991), and *Death Becomes Her* (1992), as well as the numerous science fiction and fantasy films that followed in the 1990s.[19]

The "sensation" created by morphing is not only that of the visual thrill of seeing one entity magically, seamlessly, and impossibly transformed into another. For the producer of the image, that is a matter of technology and technique. For the viewer, there is a thrill evoked by the sense of boundary disintegration, in which identity and existence are not constricted by the peculiarities and limitations of the flesh or logic or reality. As the film *Terminator 2* showed, the cyborg body, itself a morph of human and technology, was capable of becoming floors, knives, liquid metal, and even convincing humans—as if all of these were mere equivalents. Morphing breaks expectations and rules, violates "natural law," and, like shamanistic metamorphosis, can open the door on a different layer of life. It is thrilling to see what a morphing program can conjure next, what else it can (re)figure and (re)animate, and perhaps most intriguing, what taboos it can violate.

Morphing, however, does not quite qualify as the current incarnation of age-old ideas of metamorphosis and shamanic logic. What is missing to make this match is a sense of morphing in its cultural context and the recognition of the potential of morphing logic to shake up the world as we know it. This connection to the cultural context of morphing is curiously absent from the many books, magazine articles, and television specials that attempt to explain computer-generated imagery to a general audience. "How did they do that?"[20] is the question framing these metatexts as we are shown technicians creating effects that are "difficult, very difficult, most difficult and miraculous."[21] What is not asked is "How and what does it mean?"

To understand what significance morphing might have for our culture, we can consider what metamorphosis signified for the Olmec of ancient Mesoamerica who thoroughly incorporated shamanic transformation into their cultural fabric and social structure. Beginning more than three thousand years ago, from approximately 1200 B.C. to 400 B.C., a civilization known as the Olmec flourished in what is now central Mexico, Guatemala, Honduras, and Costa Rica.[22] The first complex culture in the New World, the Olmec established the basic form of a pan-Mesoamerican cultural complex that was to later include the Aztec, Toltec, Teotihuacano, Zapotec, and Maya. The Olmec lived in a complex environment, "in the very hot, very wet tropical lowland . . . amid a tangle of rivers, swamps, high jungle and savannas."[23] The Olmec are known through their archaeological remains (which include monumental architecture, pottery, and stone tools), their monumental art (most notably large carved heads), and their portable art, which is often found in the tombs of rulers. Olmec art is especially important in Mesoamerican studies because "the religious and political systems expressed through Olmec iconography established the philosophical and political foundations for successive Mesoamerican civilizations until the coming of the Spanish in the sixteenth century."[24] There are no written records of this culture, so the art has been studied as "the material expression of the concerns and beliefs of a local society interpreted through the mandates of its political leaders, the artist's individual predilection, and the Olmec ideological system."[25] Shamanic activities emphasizing metamorphosis have been widely confirmed in this and most other Mesoamerican cultures.

Olmec art focuses mostly on the human form and includes monumental sculptures (colossal heads), smaller sculptures, stelae (erect markers placed in front of buildings), masks, thrones, and figurines. The human figures depict "well-fed infants, stocky adults, or wrinkled old people" with almond-shaped eyes and downturned mouths; best-known after the colossal heads (most often thought to be ruler "portraits") are the baby figures with enlarged and elongated heads, and pudgy limbs, probably representing royal offspring or infant sacrifices.[26] There are also dwarfs, hunchbacks, deformed fetuses, contorted shamans, and other figures related to the "narrative subjects" of Olmec art: shamanic transformation and rulership.[27] Of special interest here are the Olmec sculptures and figurines that "often portray composite beings that are biologically impossible, mingling human traits with characteristics of various animals."[28] These depict steps in the sequence of metamorphosis between a shaman and his animal counterpart as he moves into the other worlds. The other human figures also represent stages of ritual action, especially those associated with agriculture and fertility and renewal. The jaguar is a motif that recurs in all these forms, even in the colossal heads, several of which seem to be wearing jaguar masks.[29] "When one attempts to classify Olmec human figures," says Ignacio Bernal, "without realizing it one passes to jaguar figures."[30] This smooth shift in typology suggests "a carnal association of humans with animals" to the Olmec.[31]

These transformation figures represent the shaman's journey in which he changes into a jaguar to navigate the otherworld successfully. They show both the agony and the ecstasy of the transformation, the power of the metaphysical experience despite the great physical pain. The jaguar is an animal protector and companion spirit (also referred to as a coessence, or *nagual*) of the shaman, whose goal is an ecstatic transformation into this alter ego. As Peter T. Furst explains, "It is precisely the sort of physically and mentally exhausting crisis—the crossing of the threshold between two worlds, two kinds of reality, if you will, two kinds of being, complementary rather than antithetical—that is integral to the practice of ecstatic shamanism everywhere, and that in the American tropics to this day often expresses itself in complete identification with, and transformation into, the most powerful predator of the rain forest and savannah."[32] Indeed, this ability to transform at will into a jaguar is a significant aspect of various belief systems in different linguistic, temporal, cultural, and geographic groups of Mesoamerica and South America.

The jaguar is more than just a major predator in these regions. It is a "metaphor for mystery and the supernatural," a symbol of power, and a model for hunting humans.[33] What makes the jaguar so appealing as an alter ego for the shaman is its capacity to cross ecological and behavioral boundaries. Whereas most other animals are restricted to specific environments and niches, the jaguar can cross many of them; whereas many animals are exclusively nocturnal or diurnal, the jaguar roams both day and night; it lives in deep jungle and swims in water as well as roaming in the open. The jaguar is an "animal that participates in various dimensions, air, land and water, and it belongs to light as well as darkness," explains Furst.[34] It is inherently supernatural and human at the same time—

the jaguar and the shaman are equivalents, and "each is at the same time the other."[35] Ethnographic analogies provide some hint of the richness of shamanic transformation into a jaguar for the Olmec.[36] Shamans were probably experienced by others literally as jaguars, as having jaguar souls and characteristics, and as being the "human colleagues" of the jaguar.[37] This jaguar soul might cause the shaman to search for his jaguar lover in the forest, paint himself like a jaguar in battle, roar, or desire meat like a wild animal. Jaguars who appeared near human settlements or attacked humans were respected as holding the souls of dead shamans. In many South American cultures, the same linguistic term is used for the jaguar and the shaman. Transformation into a jaguar was enabled by taking hallucinogenic substances, chewing coca, smoking tobacco, somersaulting or tumbling, and using snuff. In his ecstatic state, the shaman's other self, the jaguar, is revealed.[38] Were-jaguars could be killed only by magical means; if killed, they return to their former human shape (much like the werewolf in Hollywood movies).

Why, asks Furst and others, should this transformation and this equivalence between shamans and jaguars so occupy the Olmec? The activities of the shaman are not strictly "religious" and always cross over into the realm of politics. Access to the supernatural was the road to legitimacy, power, and rulership and was expressed in Olmec iconography via transformation figures. A ruler in Mesoamerica was not only a figure who wielded the power to cast judgments, dictate economics, and engage armies in battle but also the one who could use concepts from the supernatural to prove that he was the legitimate occupant of the throne.

Based on the context in which many of these transformation figures are found (buried in the tombs of elite members of society), it is likely that these Olmec figures were related to the rituals the elite were required to perform to maintain their position of power. In many ancient Mesoamerican societies, "the function of public art and architecture was to define the nature of political power and its role as a causal force in the universe."[39] As Kent Reilly explains, human rulership in Olmec times "was legitimized by a charter that stressed the ruler's access to supernatural powers," and "the Olmec-style symbol system was a visible charter for rulership."[40] The "were-jaguar transformation was a magical ability reserved to the elite . . . and the shaman-jaguar transformation played a large part, not only in the religion, but also in the confirmation of 'royal power.'"[41] For the Olmec (and as can be seen later with the Maya), "to open a portal in the invisible membrane that separated the natural and supernatural worlds . . . allowed access to a crucial source of royal power: ancestors."[42]

The populace would be familiar with these images of transformation that confirmed the ruler's right to rule, defined the social order, and described how the universe worked.[43] People "literate in the symbol system," even if they spoke different languages, could understand this public proclamation of the ruler's claim to legitimacy.[44] The shaman thus works to confirm the ruler's right to power because in the Olmec world and the Mesoamerican cultures that followed, metamorphosis was an expression of the ability to control and manipulate the material and spiritual worlds. Olmec rituals and beliefs about metamorphosis, as well as the political and economic structures they defined, laid the foundation for centuries of cultural development in a large region of the so-called New World, setting a common legacy for the native peoples of Mesoamerica.

Shape-Shifting Reality

In computer graphics, warping is the process by which image A changes into image B. In morphing, images A and B combine to produce a new image that is a hybrid of the two. Simple warped images, like the Exxon ad of the car and the tiger, show a transition, but the final product is not extraordinary; the warping software functions to eliminate differences between the two entities. Morphed images, which multiply rather than eliminate differences, provide a more chilling sense of boundary manipulation, like the shaman who is always already a jaguar. Television advertising has tended to favor the warped image, since it flattens difference so that one manufactured product can change efficiently into another. The advertising on children's Saturday morning television, however, is awash with truly morphed images, children who become one with their candies and toys. In an ad for Betty Crocker's "Fruit Gushers," children pop one of the juicy candies into their mouths and suddenly find themselves sporting a head that is a hybrid of kid and fruit, a talking lime or watermelon.

Michael Jackson's *Black or White* video (produced in 1991 by PDI) demonstrates a dynamic morph in which a series of multiracial young people change into each other, momentarily producing stunning hybrids before the morph moves on to the next racial type. This same emphasis on the power of racial blending can be seen in a 1993 issue of *Colors* magazine produced by the Benetton clothing company. In a series of morphed still images titled "What If . . . ," Queen Elizabeth and Arnold Schwarzenegger are shown as blacks, the pope is Asian, and Spike Lee and Michael Jackson are white. Less thought provoking, however, is the simple warp of a black panther into Michael Jackson in Jackson's video. Although Jackson himself, with all his physical transformations, can certainly stand as living evidence of the hybrid possibilities of morphing, this video image provides no sense of the powers of shamanic blending, where a human and a jaguar coessence inhabit each other's lives and bodies.

In one interesting example, morphing was used to create images of humans that otherwise had no other means of, or reason for, coming into existence. The cover of a magazine called *Games* (June 1993) featured a morphed image of Arnold Schwarzenegger and Roseanne Arnold that resembled neither but strangely evoked elements of both of these strong personalities. Inside the magazine was a series of images that featured more morphs of other famous people who shared one element of their name—James Dean and Dean Martin, Bob Hope and Hope Lange, John Wayne and Wayne Newton, Aretha Franklin and Franklin Roosevelt—but no other logical feature. The cover ungrammatically asked, "Who Are This?", a question that suggests the morph's capability to shake the established grammatical order.

Computer morphing has the potential to challenge the very basis of human thinking and existence—the processes by which we categorize the world. Cognitive scientist George Lakoff has described categorization as the foundation of our "thought, perception, action, and speech."[45] He explains that if we didn't categorize things, people, events, and experiences, we simply could not function socially, intellectually, or physically. It's important, for example, to be able to categorize dangerous things that are a threat to our lives, our beliefs, or our social group even though our categories are not the same as another culture's. If we can understand how a group categorizes, we can understand how they think and how they function. Categorization, emphasizes Lakoff, is "central to an understanding of what makes us human."[46]

Traditional notions of categorization create groupings whose members share some characteristics or properties. A society like ours, which is based on these traditional notions of categorization, tends to propose that there are distinct material boundaries between things, that concepts are either true or false, and that meaning is transcendent and will be generally agreed on by any reasonable person. It tends to see this reason as disembodied and abstract, "distinct on the one hand from perception and the body and culture, and on the other hand from the mechanisms of imagination, for example metaphor and mental imagery."[47] Meaning, in addition, is supposed to be objective and have reference in the real world.

Lakoff has been challenging these ideas about categorization. His revised concepts of categorization propose that categories are socially constructed and that categorization is a matter of "human experience and imagination," that categories are not set in nature, and that boundaries are fluid, with meaning contextual and subject to multiple interpretations.[48] Computer morphing can take the revision of categorization one step further. Morphing forces previously unthinkable connections and dissolves our culturally derived category boundaries—between living and dead, animal and human, inert and active, male and female, old and young. Thus computer morphing can clarify, make visible, and perhaps even accelerate conceptual change by questioning categorization altogether. As Lakoff proposes, "To change the concept of category itself is to change our understanding of the world."[49] That notions of reality, fantasy, and imagination can be shifted radically by computer-generated images (CGI) may be evident in the comments of one critic who claimed that special effects in films after *Terminator 2* would have "flights of fancy so realistic that audiences won't ever suspect they're seeing an act of industrial imagination."[50]

Even as morphing seems to be able to push us to rethink and shift the terms of our existence, we still live mostly according to our previously defined culturally bound categories. Ironically, this is most evident in the explanations and descriptions of morphing used by the industry that creates most of these images. Morphing producers revel in their ability to use this transformational technology primarily to simulate the realism of the established world. The artificial simulation of reality has been the goal of CGI (which includes morphing) since its inception. The SIGGRAPH trade show/technical meeting (a Special Interest Group for computer graphics that is part of the Association for Computing Machinery) is the premier annual conference where the technologies and techniques of computer graphics are displayed and debated. Recently SIGGRAPH has shifted away from academic and defense applications of computer imaging to applications in the world of entertainment and motion pictures. Continuing, however, has been the SIGGRAPH tradition of promoting a convincing reproduction of the external world as the goal of its activities: "For many at SIGGRAPH, the merit of duplicating the appearance of the physical world with computer imaging technology is unquestioned. Through the years the development of techniques to create 'realistic' imagery—recreating the pictorial sensibility of the Renaissance—has dominated the research papers at SIGGRAPH."[51]

When realism is the goal, anything that detracts from the illusion of reality is seen as a mistake. In computer graphics this mistake is seen as a historical problem, one based on the limitations of technology at particular points in time. However, "the subjectivity of the recognition of the real is hard to ignore.... Pictures hailed as 'realistic' a few years ago are now looked on with patronizing affection as the best that was possible at the time."[52] At this point in morphing's history, we should be ready to move from the concerns of realism.[53] The reality question, philosopher of technology Michael Heim reminds us, "has always been a question about direction, about focus, about what we should acknowledge and be concerned with," not about some fixed definition of "the real."[54] Clearly, entertainment CGI has already shifted away from the notion of a physical, scientific reality, the "eternally fixed essence of things" that was required for scientific visualization and military applications.[55] After all, most CGI is fantastic, presenting creatures that have no real-world referents and live in places ungoverned by the laws of physics, biology, and gravity. To continue to define its concerns as an attempt to convey "realism" is misleading. Yet this way of seeing is even built into the software used to create CGI. As Kopra explains, "All commercial computer animation systems obligingly simulate single-point perspective in the way they model."[56]

What should morphing be concerned with? Whereas the "invention" of linear perspective in art has given us a systematic sense of space, morphing's visual strategy should be more consistent with those artists who "since the eighteenth century have been fascinated by... deviations from strict perspective."[57] The purpose of these deviations, in, for example, surrealist art or cubism, was to break up "the ancient system's fixed, unitary, hierarchical focus."[58] For cubism this meant defining "democratically multiple perspectives," "mixed or composite images," and "a sense of a new, fourth dimension: time."[59] For surrealism it meant acknowledging that there are "other and more profitable ways of responding to an image than those accessible to reason."[60] For morphing it means foregrounding its one, powerful, and magical element: the ability, like that of the Olmec shaman, to move laterally across categories to gather knowledge from each place of (however temporary) residence. Describing the power of the dynamic morph in Michael Jackson's *Black or White* video, Anderson comments: "This is a terrific use of morphing—not just as a special effect, but a startling, visual renunciation of racism.... Thousands of years of written arguments are bested by this little video with its profound visual impact."[61] It is not that a new combination human is created and racism is eliminated, but rather that the changes between us are shown as never ending and the state of human/racial existence is always in flux. Movement is lateral rather than hierarchical.

It is very likely that computer effects are "subtly changing the nature of reality as experienced through moving images."[62] A culture based on believability and credibility has to meet different criteria for its stories, myths, rituals, and representations. Reality testing can give way to "unreality-testing" in which we find a way to reconcile personal perceptions with the corroborating witnessing of others in our culture.[63] The Olmec shaman, like ritual specialists in magical transformation everywhere, relies on the affirmation of his audience and the effects of his journeys to confirm his metamorphosis. Shared images and stories provide knowledge beyond their immediate content, and because they require consorting with others, they usually offer "profound insight into human life of a moral, metaphysical, symbolic, religious, unconscious, and narrative nature."[64] These images and stories are attempts at comprehension that are usually more useful, often more convincing, and even more "alive" than the guidelines provided by logic, realism, and rationalism. "In order to make up our minds, we must know how we feel about things; and to know how we feel about things," explains anthropologist Clifford Geertz, " we need the public images of sentiment that only ritual, myth, and art can provide."[65]

Morphing's End

Even as his colleagues were describing more precise code for creating this effect, one producer of digital imagery at SIGGRAPH 1997 declared morphing obsolete. But morphing is not just a trend in the culture of digital visualization. It seems, instead, to be a marker of our time, a time in which morphlike activities and events are overwhelming our sense of boundaries, space, direction, identity, and time. Computer-generated morphing effects are no longer a specialized or surprising form of image manipulation: through its appearances in numerous advertisements, music videos, and movies, morphing has entered the vocabulary of the masses and has shifted the grammar of our culture, a grammar that allows and enables us to think, act, experience, and categorize in particular ways.

Morphing is one of our leading tropes at the end of the century, a visual figure of speech in which one thing is presented as another, in which several entities lose themselves as they combine to create new elements and new meanings. Like verbal metaphors that involve relating seemingly disparate entities, this visual strategy creates new possibilities of seeing, thinking, and feeling. At a time and in a culture that is grappling with the ancient problems of social disjuncture, boundary disintegrations, uncertainty and ambiguity in identity, and changes in notions of space and time, morphing, like shamanic metamorphosis, is a visual demonstration of the boundary fluctuations that humans and their worlds, natural and manufactured, are experiencing. By materializing these profound shifts in the visual representation of a morph, and by foregrounding the morph as an appropriate metaphor, our culture is saying something not just about the technologies that make such imagery possible but also about the fears and fantasies that such technologies and their effects produce.

Notes

Many thanks to Michael Blitz for his ongoing discussions about the significance of morphing.

1. The exhibit, displayed at the National Gallery of Art, Washington, D.C., in 1996, was called "Olmec Art of Ancient Mexico."

2. The statues were not found together and now reside in different collections: the Los Angeles County Museum of Art, Princeton University, and Dumbarton Oaks Research Library and Collections.

3. F. Kent Reilly, "The Shaman in Transformation Pose: A Study of the Theme of Rulership in Olmec Art," *Record of the Art Museum of Princeton University* 48, no. 2 (1989): 12.

4. Ibid., 9.

5. Christopher W. Baker, *How Did They Do It? Computer Illusion in Film and Television* (Indianapolis: Alpha Books, 1994), 119.

6. See a recent discovery of cave art in Jean-Marie Chauvet, Éliette Brunel Deschamps, and Christian Hillaire, *Dawn of Art: The Chauvet Cave* (New York: Harry N. Abrams, 1996).

7. The Art Museum, Princeton University, ed., *The Olmec World: Ritual and Rulership* (New York: Harry N. Abrams, 1995), 126.

8. Leonard Barkan, *The Gods Made Flesh: Metamorphosis and the Pursuit of Paganism* (New Haven: Yale University Press, 1986), 17.

9. Ibid., 18.

10. Anne Hollander, *Moving Pictures* (New York: Alfred A. Knopf, 1989), 5.

11. Piers Vitebsky, *The Shaman* (New York: Little, Brown, 1995).

12. Nevill Drury, *The Elements of Shamanism* (Rockport, Mass.: Element Books, 1989), 1.

13. Ibid., 6.

14. Peter T. Furst, "West Mexican Tomb Sculpture as Evidence for Shamanism in Prehispanic Mesoamerica," *Antropologica* 15 (1965): 75.

15. Vitebsky, 17.

16. Ibid., 15.

17. The technique was first shown publicly at SIGGRAPH in 1982.

18. Mark Cotta Vaz and Patricia Rose Duignan, *Industrial Light and Magic: Into the Digital Realm* (New York: Ballantine Books, 1996), 114.

19. Digital morphing is now a technique available to anyone with a desktop computer using free or inexpensive software.

20. For examples see Baker, *How Did They Do It;* John Clark, "Shot by Shot," *Premiere* (August 1994): 60–62; Aljean Harmetz, "Two Special Effects (a Crib Sheet)," *New York Times,* 24 July 1994, 13, 24; Charles Solomon, "How They Did That," *Los Angeles Times,* 22 September 1991, 3, 39–40.

21. Vaz and Duignan, 204.

22. Mesoamerican scholars disagree about whether the term *Olmec* should apply to a distinctive culture and people, or should simply refer to a widespread art style. See discussions in *The Olmec World: Ritual and Rulership* and in Elizabeth P. Benson and Beatriz de la Fuente, eds., *Olmec Art of Ancient Mexico* (New York: Harry N. Abrams, 1996).

23. Michael D. Coe, *America's First Civilization: Discovering the Olmec* (New York: American Heritage, 1968), 7.

24. Reilly, 5.

25. Carolyn E. Tate, "Art in Olmec Culture," in *The Olmec World: Ritual and Rulership,* 49.

26. Tate, 56; also see Peter David Joralemon, "In Search of the Olmec Cosmos: Reconstructing the World View of Mexico's First Civilization," in *Olmec Art of Ancient Mexico,* 51; and the National Gallery show brochure, Susan M. Arensberg, "Olmec: Art of Ancient Mexico" (Washington, D.C.: National Gallery of Art, 1996).

27. *The Olmec World: Ritual and Rulership,* 126; Tate, 56.

28. Joralemon, 51.

29. Ignacio Bernal, *The Olmec World,* trans. Doris Heyden and Fernando Horcasitas (Berkeley: University of California Press, 1969), 57.

30. Ibid., 66.

31. Ibid., 67.

32. Peter T. Furst, "Shamanism, Transformation, and Olmec Art," in *The Olmec World: Ritual and Rulership*, 70.

33. Benson and de la Fuente, 228.

34. Furst, "Shamanism, Transformation, and Olmec Art," 75.

35. Peter T. Furst, "The Olmec Were-Jaguar Motif in the Light of Ethnographic Reality," in *Dumbarton Oaks Conference on the Olmec*, ed. Elizabeth P. Benson (Washington, D.C.: Dumbarton Oaks Research Library and Collection, 1968), 148.

36. Ibid.

37. Ibid., 158.

38. Ibid., 163.

39. Linda Schele and Mary Ellen Miller, *The Blood of Kings: Dynasty and Ritual in Maya Art* (New York: George Braziller, 1986), 103.

40. Reilly, 5–6.

41. Ibid., 15.

42. Ibid., 16.

43. Schele and Miller, 103.

44. Reilly, 6.

45. George Lakoff, *Women, Fire, and Dangerous Things: What Categories Reveal about the Mind* (Chicago: University of Chicago Press, 1987), 5.

46. Ibid., 6.

47. Ibid., 7.

48. Ibid., 6.

49. Ibid., 9.

50. Guy Garcia, "Make Sticky, Morph!" *Time*, 8 July 1991, 56.

51. Andy Kopra, "Dinosaurs and Dialectics: Computer Graphics at SIGGRAPH '93," *New Art Examiner* (December 1993): 22.

52. Ibid.

53. By 1997, SIGGRAPH participants were exhorting their colleagues to create effects that went beyond attempts at realism, that instead were believable, enjoyable, and captivating, and that drew the audience into alternative immersive worlds of narrative and empathetic characters.

54. Michael Heim, *The Metaphysics of Virtual Reality* (New York: Oxford University Press, 1993), 118.

55. Ibid., 133.

56. Kopra, 22.

57. Scott Anderson, *Morphing Magic* (Carmel, Ind.: Sams Publishing, 1993), 108.

58. H. H. Arnason, *History of Modern Art: Painting, Sculpture, Architecture, Photography,* 3d. ed. (New York: Harry N. Abrams, 1986), 142.

59. Ibid.

60. J. H. Matthews, *The Imagery of Surrealism* (Syracuse, N.Y.: Syracuse University Press, 1977), 4.

61. Anderson, 232.

62. Woody Hochswender, "When Seeing Cannot Be Believing," *New York Times,* 23 June 1992, 1.

63. Wendy Doniger O'Flaherty, *Dreams, Illusion, and Other Realities* (Chicago: University of Chicago Press, 1984), 175.

64. Charlotte F. Otten, ed., *A Lycanthropy Reader: Werewolves in Western Culture* (Syracuse, N.Y.: Syracuse University Press, 1986), 223.

65. Clifford Geertz, *The Interpretation of Cultures* (New York: Basic Books, 1973), 82.

4

Cultural Transformations from Greek Myth **From Mutation to Morphing** to Children's Media Culture

MARSHA KINDER

> The possession of originality cannot make an artist unconventional; it drives him further into convention, obeying the law of the art itself, which seeks constantly to reshape itself from its own depths, and which works through geniuses for its metamorphoses, as it works through minor talents for mutation.
>
> **Northrop Frye**

> We are more willing to act in the U.S. like a U.S. company, in Europe like a European company, and in Japan like a Japanese company. That's the only way a global company like Sony can truly become a significant player in each of the world's major markets.
>
> **Akio Morita, founding chairman of Sony**

Origins

These two epigraphs (recycled from two of my earlier works) function as launchpad for this essay on mutation and morphing, propelling me both backward with Frye in search of mythic roots for postmodernist shapeshifters and forward with Sony to reveal strategies of survival for marketing geniuses who increasingly replace artists in the global economy of the 1980s and 1990s. The meanings of these quotes will be transformed within this new discursive context—a study of the recent meteoric rise of this pair of tropes (mutation and morphing) within children's media culture.

More specifically, I will trace how these images of cultural transformation have been mythologized in those two controversial bands of shape-shifting superheroes, the Teenage Mutant Ninja Turtles, who rose to global cult status in the mid-1980s, and the Mighty Morphin' Power Rangers, who displaced the Turtles in the 1990s.

Both of these cults built on the earlier success of the toy genre known as "Transformers," a species of action figure that enables young owners, with minor deft twists, to convert a formidable creature (robot, monster, or superhero) into a high-powered vehicle or weapon, and vice versa.[1] The Teenage Mutant Ninja Turtles made their pop culture debut in 1984 in a limited-edition comic book designed by two young unknown American artists, Peter Laird and Kevin Eastman, who thought they were parodying superheroes. But the myth and marketing of these transformers soon captured the attention of children and the media first in the United States and then worldwide, generating an intense "Turtlemania" that spawned a proliferating catalog of licensed products, including action figures, plush toys, pajamas, T-shirts, Halloween costumes, cereals, juices, pizza, lunch boxes, martial arts schools, party supplies, board games, toy weapons, video cassettes, books, calendars, and talking toothbrushes. You name it, they licensed it! The phenomenal selling power of this cult began to be fully registered by the media in 1987 when despite a decline in the sales of boys' action figures, Playmates introduced a new set of collectible TMNT action figures, which quickly sold out in toy stores across America. By the 1990 Christmas season, there were forty-four of these Playmate action figures on the market. Not only were the Turtles impressive transformers within their narrative domain (miraculously mutating into giant, talking, pizza-loving amphibians after having been exposed to a mysterious toxic goo in an urban sewer and then quickly being transformed into martial arts experts by their Japanese rat guru, Master Splinter), but they also were fabulously successful in fluidly moving from one mass medium to another, carrying not only Playmates but also their other associates along for the lucrative ride. Their hour-long animated TV show, which premiered in the fall of 1990 on CBS, quickly made that station the top-rated network in Saturday morning television.

When the Hong Kong distributor Golden Harvest released the first TMNT movie in 1990, it took in over $25 million the opening weekend (at that time one of the biggest-grossing three-day openings in history). The Turtles myth scored similar successes in videotape rentals, home video game sales, and arcade games, reaching their commercial peak in 1991, the year my own book about them was published and shortly after the 1990 passing of the Children's TV Act, which led to a closer critical scrutiny of children's programming. Although you can still find the Turtles in the land of TV syndication, by the May 1994 sweeps, they were off the charts and their sales in licensed products had seriously slipped. Yet between 1984 and 1994, more than $7 billion of Turtlized merchandise was sold. Ten years isn't bad for any buying bonanza, particularly in children's pop culture, where few besides Barbie survive.

Carefully scrutinizing the phenomenal international success of these mutant transformers was a young, Egyptian-born Israeli independent producer named Haim Saban, who emigrated to Hollywood in 1985, only one year after the Turtles' debut. According to his banking agent John Shuman, Saban's driving ambition was to become "a full-service software provider" with a "plan . . . clearly calculated at exploiting the children's market internationally."[2] What Saban found in the Turtles was a successful formula for a cultural myth that could enhance a global marketing strategy—a combination of pan-Asian martial arts action with a wholesome gang of heroic American teens. In other words, a comic conflation of Hong Kong and Hollywood superheroes to colonize the world! Buying the rights to a Japanese action series for children (which was already popular in that nation) and adapting it to kids media culture in the States, he hoped to make it marketable worldwide, which is precisely what he succeeded in doing with *The Mighty Morphin' Power Rangers,* which debuted on American television in 1993. Transforming the Japanese live-action heroes into a band of ordinary American high school kids, he diversified their identities in terms of gender and ethnicity while retaining the Japanese action footage (where they are garbed in unisex jumpsuits, helmets, and boots that reveal no flesh or nationality). He also retained the talking head of their Master Zordon, as well as their Japanese antagonists, the villainous Rita Repulsa and her legions of invasive aliens. Thus despite the tropic shift from mutation to morphing in his heroes, on the levels of both myth and marketing, Saban's hybrid creation (in Frye's terms) was more like a minor mutation than a true metamorphosis.

Yet like their amphibious precursors, the Rangers also transported their power brokers to new heights—especially Saban himself, who, according to Mike Freeman, "is one of the few truly independent independents."[3] Despite the multinational diversity of his backers and the bonding among his band of heroes, when Saban uses the term "we" (unlike Sony's Akio Morita), he is usually referring to himself. Thus his profits from the Power Rangers could surpass even those of the Turtles, even if their bonanza doesn't survive a full decade like that of their precursors. With the long run of the Turtles clearly in mind, even as early as 1993 Saban was calling the Power Rangers a "ten-plus-year, multibillion dollar franchise. . . . Our whole approach—our investments, our expectations for the return on investment—is based on a ten-year plan and not a two-year plan."[4] By December of that year, Freeman reported: "Nearing the close of the Christmas sales season, sole toy licensee Bandai Co. has sold close to one million of the *Power Ranger* action figures, putting it on a faster early pace than the record sales of *Teenage Mutant Ninja Turtles* dolls five years ago."[5]

The other two power brokers who profited from the meteoric rise of the Rangers were Fox TV (which was then struggling to join ABC, CBS, and NBC as one of the top national networks) and the new president of Fox's Children's Network, Margaret Loesch. Frequently called "the Queen of Kidvid," Loesch (before joining Fox in 1990) was at NBC and Hanna-Barbera, where she did *The Smurfs,* and then moved to Marvel, where she developed successful television series such as *G.I. Joe, Transformers, My Little Pony,* and *Muppet Babies.* Being a newcomer at Fox, she still had difficulty persuading her new colleagues to trust her instincts. In a "Special Report on Children's Television" for *Mediaweek,* Eric Schmuckler claims that "Loesch relishes telling how she fought to put the show on against the judgment of her staff, bosses, and affiliates."[6] In fall 1993, just three weeks after its premiere on Fox on weekdays at 7:30 A.M. (generally considered a "throwaway" time slot), *Power Rangers* captured first place in the ratings and remained number one (with viewers under age eighteen) for the entire season. With younger kids from two through eleven, the show had an amazing 52 percent share. This new popularity helped move the show to the privileged Saturday morning lineup, where it almost immediately took first place and helped the Fox Children's Network become (after only four years on the air) number one in the $650 million kids TV market. After a comparable four years of success for the Rangers, their popularity has already begun to subside. Unlike that of the Turtles, their movie (released in 1995) was a big disappointment, and it is now doubtful whether they will fulfill the ten-year plan projected by Saban. But at the time of this writing (August 1997), the latest incarnation of their series, *Power Rangers Turbo* (which recharges their selling power by driving them back to their roots in the Transformer toy genre), is still playing on Saturday morning television.

Thus far I have been talking mainly about marketing and saying very little about the mythic appeal of the Turtles and Power Rangers or the relationship of that appeal to the transformative tropes of mutation and morphing. Although the popularity of both sets of superheroes is now in decline, they have entered the popular imagination of global culture as an optimistic myth of comic transformation not just for kids but for cultural theorists talking about the reproduction of postmodernist subjectivity. For example, if you turn to the home page of cyberspace theorist Sherry Turkle, you can download a video that shows the Zordon-like bald head of Michel Foucault morphing into a Power Ranger, implying, as Edward Rothstein puts it, that "the morphing of the philosopher into a pop figure may support Foucault's argument that identity is elastic."[7] This argument is consistent with one I made in *Playing with Power,* which read the myth of the Mutant Turtles as a global force that contributes to the mass reproduction of postmodernist subjectivity.

> Evoking the comic prototype of Proteus (the seagod who fluidly changes shape), the Turtles' . . . status as amphibians, teenagers, mutants, and American ninjas with Italian names and California surfer jargon quadruples their capacity as transformers, making them the ultimate sliding signifiers: they can easily move from an animated TV series into a live-action movie, and they can transgress borders of species, race, ethnicity, generation, and media. While such cross-cultural malleability might help construct subjects who are less prejudiced against alien Others, the changes promoted are far from revolutionary.[8]

Before we can fully address what the Mutant Turtles and Power Rangers have in common or, for that matter, how they differ (as mutants or morphs), we need briefly to explore what was at stake in some of these earlier representations of shape-shifters, for only then can we determine in what sense the Turtles and Rangers represent a cultural innovation.

Mythic Roots

Metamorphosis is a trope that is central to creation myths from many cultures, where it frequently serves as an image of creation or destruction, reward or punishment, growth or decay, or the passage from life to death. It is also a defining formalist feature of dreams and their characteristic tropes of condensation and displacement, where the mere temporal or spatial proximity of two juxtaposed images can, when narrativized, be read as transformative change—a cognitive process that is fundamental to flip books, surrealist jolts, trick films, animation, the basic illusion of cinema, and the visual perception of movement.

Although we can find shape-shifters in myths from most cultures, a complete genealogy of such figures is far beyond the scope of this essay.[9] What I intend to do here is merely suggest some of the issues at stake in a few of the best known of these "transformers" from earlier eras in Western culture, including Proteus, the Egyptian sea god colonized by the Greeks; Tiresias, the Greek prophet; and Morpheus, the Roman god of dreams. Although several Greek and Roman goddesses (such as Athena and Hera, as well as several female figures in Ovid's *Metamorphoses*) were also capable of shape-shifting, significantly it was male figures in these cultures who were primarily identified with these powers.

In contrast to the Turtles and the Power Rangers, who star in a string of serial adventure narratives that celebrate their ingenuity and ability to survive, singular classical shape-shifters such as Proteus, Tiresias, and Morpheus played only supporting roles. Certainly none of them was an action hero wielding a sword like Achilles or outwitting his enemies like Odysseus. In fact, they didn't benefit directly from their own supernatural powers. Nor did they help the creative power brokers (like Homer, Sophocles, and Ovid) who depicted them (the way the Turtles empowered Eastman and Laird, and the Rangers Saban and Loesch). Nor did the classical shape-shifters enhance the networked powers of the supreme overlords they served (such as Poseidon and Zeus, the way our modern shifters elevated CBS and FOX). Instead, they functioned merely as mediums, each with his own favorite vehicle (magic, prophecy, or dreams) for narrowcasting divine data to human petitioners so that these users could enhance *their* powers of survival. In other words (or in Proppian terms), these shifters were helpers rather than heroes.

Proteus appears in book 4 of Homer's *Odyssey* (c. 850 B.C.), where he is described by his daughter Eidothea as "the Egyptian, immortal Proteus, [who] knows all the depths of the sea, [and] is Poseidon's underling," and whose magic arts enable him to take the form of "all creatures that come forth and move on the earth, he will be water and magical fire."[10] He uses these dazzling transformations (into a lion, serpent, leopard, boar, fluid water, and tree with towering branches) not to gain knowledge but rather to frighten and evade those who want access to his divine data.

Proteus's human counterpart was Tiresias, the blind Theban seer, who also possessed these combined powers of divine knowledge and physical transformation. In one mythic version of his back story, in punishment for witnessing the coupling of a pair of snakes, he was temporarily transformed into a woman. (Undoubtedly a sex change in the opposite direction, from woman to man, would not have been considered a punishment.) Because of the special knowledge this hermaphroditic experience provided (a connection that is operative for shamans in many different cultures), Tiresias was later asked by Zeus and Hera to adjudicate their dispute over which gender experiences greater pleasure during sex. After responding that women's pleasure is nine times greater than men's, Tiresias was punished by Hera with blindness, whereby Zeus, in divine compensation, granted him the gift of prophecy. Hence, second sex metamorphosed into second sight.[11] In Sophoclean tragedies like *Oedipus* (427 B.C.) and *Antigone* (442 B.C.), Tiresias's shape-shifting is not even mentioned. In the former he is described by the Chorus as "the godly prophet . . . in whom alone of mankind truth is native," and by Oedipus as the "alone one that can rescue us."[12] Yet his special knowledge does exactly the reverse: it reveals and thereby helps fulfill the full horror of the tragedy. Tiresias functions as a divining mirror in which defiant royal petitioners can read their own cruel destinies imposed by punishing gods.[13] Whereas Tiresias confirms the tragic fates of arrogant kings, Proteus performs his transformations within the essentially comic world of the epic, which celebrates the miraculous survival powers of its hero Odysseus. It is hardly surprising, then, that Homer would emphasize Proteus's powers of shape-shifting rather than those of seeing (as in the case of Tiresias).

Like Proteus, Morpheus, the Roman god of dreams, usually performs within a comic world where the emphasis is similarly on shape-shifting rather than on prophetic knowledge. In fact, his very name derives from his ability fluidly to assume the form of any being within his virtual realm of dreams, an impersonation that frequently brings about a transformation in the dreamer. He is quite literally, then, a medium of representation. Although dreams (like prophecy) are a source of hidden knowledge (especially about the past and future), they are unique in being the prototypical medium of transformation for the individual, the culture, and the species, one that mediates between biological programming and cultural imprinting and that formalistically features visual quick-changes even more fluid than those in contemporary digital morphs.

One of the fictional worlds inhabited by Morpheus is that of Ovid's *Metamorphoses* (first century A.D.), which, despite the potential sadness of its mythic tales of rape, murder, and unrequited love, still generates a comic spirit of resilience and wit primarily through physical transformations.

Such morphs link not only his stories but also the fates of his victors and victims, who experience them either as punishment or as compensatory transcendence. For example, in Ovid's tale of Ceyx and Alcyone, Morpheus assumes the form of the drowned husband who appears in his wife's dream to tell her of his death. Determined to join him, the grieving Alcyone rushes to the shore the next morning, where she finds the corpse of Ceyx. Instead of drowning, she is transformed by the gods into a bird and joined in flight by her husband, who has undergone a similar metamorphosis. Thus this potentially tragic story is transformed by Ovid into a delightful flight of fantasy on eternal love.

These few examples suggest that classical tropes of shape-shifting were contextualized by genre and that the comic/tragic divide was definitive. In comedy, morphing served as a primary mode of survival and flexibility, both privileged values. Here transformations took many forms and were quick, easy, and reversible, as they are in the essentially comic myths of our own Mutant Turtles and Power Rangers. Conversely, tragedy demanded commitment to a single shape or body of values. Even if a tragic hero doubted the existence of absolutes, he or she could prove the absolute value of an object of worship (whether it be love, honor, duty, and so forth) through a willingness to sacrifice everything else to preserve it—by acting *as if* it were absolute, even if that meant having to die for this simulation. That is one reason why tragic transformations had to be painful and irreversible.

If Proteus was the archetype of comic transformation with his exuberant catalog of serial identities, then Narcissus (rather than Tiresias) was his tragic antithesis, stubbornly committed to his singular conversion into an unchanging imaginary reflection. This pair evokes not only the binary of comedy and tragedy but also its modern equivalent, sadism and masochism (at least as theorized by Deleuze)[14]—with promiscuous protean substitutions of frightening, malleable bodies controlled by one who is all-knowing pitted against a beautiful narcissistic image of perfectly mirror-matched lovers forever suspended within an intimate specular embrace. However, within this classical system there is still some leeway for mobility, for as we have seen, a comic protean shifter can be transformed into a tragic seer (like Tiresias in *Oedipus Rex, Antigone,* and *The Bacchae*), and the tragic fates of doomed lovers (like Pyramus and Thisbe) can be softened with posthumous transformations in Ovid's comic *Metamorphoses.*

This mobility becomes much more fluid during the Renaissance, as can be seen in Shakespearean theater, particularly in a comedy such as

A Midsummer Night's Dream (1600), where potentially tragic lovers are foolishly yearning for faery changelings and asinine shifters and where a parodic restaging of Ovid's *Pyramus and Thisbe* (with workmen performing protean impersonations of a lion, wall, and moon) becomes hilarious. Such mobility can also be found in tragedies such as *Othello* (1622), *Lear* (1608), and *Richard III* (1597), where mercurial comic characters are transformed into defiant, duplicitous villains (Iago, Edmund, and Richard, respectively). And it is even possible to read *Othello* in light of *A Midsummer Night's Dream,* particularly when Theseus observes the similarities in the transformative powers of lovers, madmen, and poets.[15]

Whereas the destinies revealed through classical mediums like Tiresias and Proteus are part of a divine system of justice—with its punishments, rewards, and retributions—those in Shakespearean tragedy are frequently improvised by inventive villains who are dissatisfied with their own subordinate position in the prevailing structure. During the Renaissance, transformative tropes increasingly take on the trace of social and political mobility that threatens existing boundaries of class as well as those of genre. This dynamic is certainly not restricted to Shakespeare but can be found in many Renaissance works structured around tropes of transformation. Two such works proved pivotal for the development of modern politics and the modern novel: Machiavelli's *The Prince* (1513), a handbook on how to turn shape-shifting into a statesman's weapon in his quest for power, and Cervantes's *Don Quixote* (1605), where not only windmills are transformed into giants and an aged Renaissance reader into a valiant medieval knight-errant but also a parodic romance into a new mode of fiction capable of dialogizing tragic and comic perspectives.

Kafka's *Metamorphosis* (1916) inverts both of these legacies. Gregor, the passive, guilt-ridden protagonist of this absurd modernist novel, is a shape-shifter who awakens one morning and simply finds himself transformed into a gigantic insect. As evasive as Proteus's metamorphoses, this grotesque transformation enables Gregor to escape the tedious, dehumanizing job by which he supports his oppressive family and the petty social ambitions imposed on him by his bourgeois milieu. This transformation functions as an "as if" simulation, not (as in classical tragedy) for an absolute value one could die *for* but a painful modernist subjectivity one could die *from.*

By the end of the nineteenth century, transformative tropes had acquired formidable new powers as driving theoretical engines of subversion, particularly in the work of Freud, where they become central to the language of dreams and its dreamwork codes of condensation and displacement, and to primary process thinking and the language of subjectivity, with its endless chain of substitutions, all of which undermine the authority of rational thought; in the work of Darwin, where they become naturalized as the primary mechanisms of evolution, which downgrade the ontological status of the rival biblical account from truth to myth; and in the work of Marx, where they become revolutionary change in the service of history, requiring ruptures that are violent and irreversible and transformations that punish and destroy obsolete systems. All these engines help drive the subversive transformation of Kafka's protagonist, which through a crisis of hysteria and subjectivity occurs as a sudden rupture that violates all rational authority of the prevailing social order and even reverses the evolution of the species. Although his radical metamorphosis leads to his own injury, imprisonment, paralysis, and death, Gregor nevertheless succeeds in evading the banal happy ending that awaits the rest of his family. In this modernist tale of masochism and horror, the fatal transformation is *both* punishment and ironic compensation.

Gregor's metamorphosis helps prefigure the legions of self-willed insectival transformations that infest William Burroughs's *Naked Lunch* (1959), changes that are shorn of masochism and guilt but are even more grotesque than Kafka's with their comic obscenities and sadistic excesses. The exuberant outpourings of a junkie's subjectivity, this dystopic "word horde" presents a satiric vision of American corruption, "a frozen moment when everyone sees what is on the end of every fork."[16] What is exposed is "a basic formula of 'evil' virus: *The Algebra of Need,*" which drives not only drug addiction and erotic desire but also the capitalist lust for money and power.[17] In this fluid world of constant metamorphosis, where naked creatures called Mugwumps "secrete an addicting fluid from their erect penises which prolongs life by slowing metabolism" and "the Dream Police disintegrate in globs of rotten ectoplasm,"[18] dependent drug addicts play parodic tragic heroes, always longing for a high-protean quick-change to evade interrogators and to disavow their commitment to a substance that requires a daily sacrifice of all other values. It is a world where everything liquefies into a protoplasmic ooze that proves more malleable than Power Goo. Where all bones, flesh, and bodies, all rooms, cities, nations, and languages, all sentences, paragraphs, stories, dreams, and genres morph promiscuously into each other, dissolving into dystopic software for a nonelectronic cyberspace.

These transformational tropes of drug experience underwent a dramatic conversion in the sixties, when they pulsed between utopic rushes and dystopic bummers, responsive to each new substance as it came onto the market. They helped spawn a new species of multisensory imagery that was widely used to describe and link different types of music and sexual experiences. Mind-expanding hallucinogens such as acid, mescaline, and peyote came to be associated with the Beatles and with the valley orgasm of Tantra yoga, while the quicksilver changes of coke and speed were linked to peak orgasms and the Rolling Stones. While Jagger was morphing into demons and outlaws in a tune like "Sympathy for the Devil," in a transformative film like *Performance* (1970), and in a violent concert like Altamont, the Beatles were acquiring new spiritual power in India and acolytes who believed unconditionally in their capacity for unlimited transformation and growth. As the Beatles became a household word worldwide, they reinscribed the insectival tropes of Kafka and Burroughs by purging them of all grotesque connotations, liberating them in strawberry fields forever and letting them take flight with Lucy in the sky with diamonds like Ovid's metamorphosed lovers. Both they and their fans had faith that no matter what medium they turned to, they would be able to create not merely lucrative ancillary products (like the Turtles and Rangers) but also new creative highs. Their power was based partly on a belief in the magical nature of their collaboration—one that combined the talents of four unique individual talents so that each experienced a radical power surge whenever they performed together. In other words, the whole was far greater than the sum of its parts, which led to a fetishizing of their bonding.

Like the Beatles, the myths of the Turtles and Rangers also promote an optimistic belief in creative collaboration and unlimited growth and also fetishize group bonding. In fact, none of these three bands has a designated leader. Unlike Elvis or even the Stones (whose headliner was first Brian Jones and then Mick Jagger) and unlike those lone superheroes from earlier decades of American pop culture, such as Superman, Batman, Spider-Man, and Wonder Woman (who sometimes had sidekicks) or Old World demonic doublers, such as vampires, werewolves, and Messrs. Hyde, the Beatles, Turtles, and Rangers all function as egalitarian members of groups who offer their fans an expressive choice in their favorite object of identification. Indeed, Loesch seemed aware of the importance of this appeal in the Power Rangers, for whenever their series came under attack, she defended it by claiming: "It's an ensemble group kids have picked up on"[19]—an argument that also applies to *X-Men* (Saban's other successful series on Fox) as well as to *TMNT.* This group ethos acquired new ideological meaning in the myths of the Turtles and Rangers, where it helped negotiate the cross-cultural gap between the American celebration of individualism and the Japanese commitment to the group.

Another similarity these postmodernist shifters shared with the Beatles is the transformative appropriation of artistic traditions from other cultures. Whereas the Turtles consciously poached the names of old masters from Italy (Michelangelo, Raphael, Donatello, and Leonardo) and martial arts from Asia and the Rangers an actual program from Japanese television, the Beatles (as well as the Stones) freely adapted African American music (as well as classical Indian music). Although the Beatles and Stones presented these borrowings as an homage, the economic dynamics were essentially the same, for the bands were the ones who profited. To make matters even more problematic, such appropriations were usually accompanied by an implicit rationalization that the "source" culture lacked the required know-how to turn these artistic assets into viable mainstream products in the world market.

In the case of African American borrowings, this appropriation was increasingly challenged, partly through a compelling demonstration of a culturally distinctive approach to tranformative tropes, which play such a crucial role in the black aesthetic on so many different registers. The richness of this cultural distinctiveness is perhaps most apparent in the improvisational structures of jazz, with its lyrical flights by legendary players such as Charlie "Bird" Parker and Louis Armstrong and innovative scat singers such as Ella Fitzgerald; in the protean multiple identities of Ralph Ellison's evasive *Invisible Man* (1952) and in the regenerative transformations of a political shifter like Malcolm X, who both used plasticity as a tactic of survival to cope with racism and the legacy of slavery; and in an artistic stylization of everyday actions such as talking, walking, dancing, and athletic moves—a stylization that generates new cultural forms like voguing and hip-hop and that transforms boxing and basketball into poetry in motion, spawning a new brand of superhero like Muhammad Ali and Michael Jordan (or their dystopic doubles, Tyson and Rodman).

The cultural specificity of these transformative tropes is brilliantly mythologized in Ishmael Reed's satiric novel *Mumbo Jumbo* (1972). Reed presents a black counterhistory of the cultural wars, tracing them all the way back to the struggle in ancient Egypt between Osiris and his censorious brother Set, the primary precursors for the opposition between the Greek gods Dionysus and Apollo. Depicting both brothers as transformers, Reed credits Osiris with being the primary spiritual source of "Jes grew," which he transmitted to Dionysus, who in turn taught the Greeks the Osirian art of revelry and dance, a tradition that runs through history and in America begets ragtime, jazz, and the Harlem Renaissance. As James Weldon Johnson is quoted as saying, "'Jes grew' songs . . . had been sung for years all through the South. The words were unprintable, but the tune was irresistible and belonged to nobody."[20] In contrast, Set is presented as the founder of the Atonists (some of whose leading advocates are Moses, Milton, Freud, and Frazer), who preach work, discipline, and militarism and who try to marshal, suppress, and exploit this spiritual heritage. Reed rewrites all other cultural histories in the process, appropriating those of the Greeks, Romans, Christians, and Jews, just as those cultures had earlier defined the African American experience. What Reed brilliantly demonstrates is the power that these transformative tropes have in the discursive wars of American popular culture—where the stakes are not simply in telling the "truth" but in presenting the most compelling version of one's own mythology. There may be no more important battlefield in these wars than that arena where conflicting versions of those myths are first transmitted to children.

Of the Species, Mutant Turtles, and Morphin' Rangers

One day while on Larchmont Avenue (a street full of posh boutiques in a white middle-class neighborhood in Los Angeles), I witnessed a fascinating scene. A young mother was walking down the street with a baby girl in a stroller and her four-year-old son by her side. Suddenly the boy jumped into a martial arts pose, shouting, "Go, go, Power Rangers!" as a look of pride and supreme pleasure spread across his face. When I got a little closer, I realized that this little flaneur was responding to a large Power Rangers poster hanging in the display window of a shop. I was intrigued with the question of why this experience was so pleasurable for him. Perhaps it was because the poster seemed to be specifically addressing *him*, rather than his mother or baby sister (who was apparently occupying more of Mom's attention than he was). Not only did the poster acknowledge his importance as a knowledgeable player within a culture, but it helped him map the world and his own place within it—a map that directly linked his own television set to shop windows on the boulevard (or, as some cultural theorists might say, domestic space and the public sphere). The poster also tested his cognitive powers, demonstrating that he was successfully able to recognize these connections and, even more important, that he knew what to do in response—he knew the correct moves and mantra and could perform them with a sense of drama and style.

These moves functioned as a form of juvenile voguing—like the kind performed in the documentary *Paris Is Burning* (1990), where adult "children" (as they are called by those interviewed on camera) fiercely compete in elaborate masquerade balls in Spanish Harlem that reassure marginalized gay men of color that they can "pass" (as a woman, straight man, white VIP, or whatever) and can feel safe as they move between "legendary houses" and public display (those protective alternative spaces of their own creation). A similar thing was happening for this boy and for other little kids in their bedrooms and playgrounds, dressed in the Power Rangers drag that empowers them to pass for teens who range more freely across different media and neighborhoods. What I had witnessed was the interpellation of a four-year-old subject into consumer culture. As part of the Power Rangers commercial network (and functioning as an Althusserian ideological state apparatus), the poster was hailing him as a postmodernist subject, and he was responding loud and clear to that call. That call and response made him feel empowered, and this feeling was the source of his pleasure.

Similar scenarios have undoubtedly been (and are still being) played out with the Turtles, for both bands of superheroes reassure children they are courageous comic survivors who can thrive even in urban sewers and public schools, despite dire threats of toxic waste, pollution, gangbangers, random violence, urban decay, and other contemporary problems that provide plots for their proliferating episodic adventures. These superheroes also give their young fans a sense of cognitive mastery, for having several transformers per set provides not only options for expressively choosing a personal favorite but also a cognitive challenge for mastering the codes that enable fans to distinguish one from another. Although the Turtles and the Rangers (once in their unisex costumes) all look the same, both sets are coded with individualizing names, colors, weapons, special powers, and personality traits that fans must memorize. Given that the Turtles carry unique "designer" labels poached from great Italian Renaissance artists, kids must learn the names of these masters and how to discriminate among their styles, but in the process they also learn how to substitute fighting and marketing for making art.

Besides the codes already listed, the Rangers also have totem animals called Zords, who function very much like astrological signs. The Power Rangers plots are usually very hard for the uninitated to follow because the pacing and transformations move faster than Roadrunner cartoons. Thus they train youngsters how to read narrative, or more specifically serial television (as well as fast-paced, cartoonish action movies like *Batman* [1989] and its proliferating sequels). This knowledge is cumulative; it rewards fans who are faithful to the show and who can therefore keep up with the twists in the plot. This time pressure (both within and across episodes) is frequently emphasized in the stories themselves, where it is usually linked to the kind of self-discipline that is essential for mastering martial arts and that the Rangers' chubby high school antagonist can never achieve.

Like their Greek precursors, both sets of superheroes also combine the twin powers of shape-shifting and knowledge. For the Rangers, their superior knowledge is displayed not only in the action sequences but also in the classroom, where it distinguishes them from the stupid, cowardly bullies who try to foil them. For the Turtles, it is mainly a knowledge of martial arts and the self-discipline that goes with it—the same lessons preached in *The Karate Kid* films. But Donatello is also singled out as the one who is a whiz at science and masterminds many of their most ingenious technological solutions.

Yet for all their know-how, in contrast to classical gods like Proteus and Morpheus, these postmodern superheroes have a fairly limited repertoire of transformations, which are restricted merely to size, fighting skills, and physical power. This limitation is not seen as a problem, since the vital issue is clearly empowerment and since these transformers achieve diversity in other ways. Like the conflict between individualism and bonding, there are many contradictions that emerge around this issue of diversity. In the case of the Turtles, who have a multinational identity despite their Italian names and Japanese master, they all speak like white California teens and are exclusively male. The only female in sight is April O'Neill, the feisty Irish American reporter who plays Lois Lane to their Supermen.

In the case of the Power Rangers, despite an earnest attempt to diversify their membership (especially in contrast to the original Japanese series, where girls were not full-fledged members and where all Rangers had the same ethnicity), the early color coding in the American series unfortunately tended to reinforce old racial and gender stereotypes: Zack, the only African American male, was the black Ranger; Kimberly, a white female, predictably pink; and Trini, an Asian American female, coded yellow. As if that were not sufficiently problematic, in the 1994 season (shortly before the Christmas shopping rush), Tommy (formerly the Green Ranger, who had originally been sent by Rita Repulsa to spy on the real Rangers but later underwent a moral conversion) was strategically metamorphosed into a White Ranger, who immediately became not only the most powerful and virtuous of the Rangers but also the new leader of this formerly egalitarian group. Not surprisingly, his action figure and other paraphernalia quickly sold out in the stores, sending droves of parents into a desperate quest for this super-desirable white hero and giving new meaning to the term "white Christmas." According to Ann Knapp, programming director for Fox's Children's Network, this change was motivated by the character's virtue ("You could look at it as Tommy being rewarded for being such a good student and working so hard on his martial arts"), yet cynics claimed, "it's because Saban Entertainment . . . ran out of Green Ranger footage from the cheap action sequences bought for the show from Japan" and "it also has to do with product turnover."[21] Far more disturbing were the racist implications, for this transformation clearly put the white man's power back on top; yet it simultaneously exposed the constructedness of "whiteness" as a category. Unfortunately, in both the Rangers and Turtles series, the only diversity that ultimately counts is the ability to move as a successful commodity across different platforms and media. For in these myths, despite all the moralistic rhetoric about fighting to save the world, the mastery of marketing proves to be more important than the mastery of martial arts or justice.

As the prime spokesman for justice, both teenage bands have an adult guru with godlike powers whose body is restricted in form. Like the blind prophet Tiresias, their special knowledge and verbal mastery are linked with a physical handicap: Zordon is a talking head without a body, and Splinter is a giant talking rat. Not only are their static corporeal forms radically different from those of their supple young protégés, but these gurus are incapable of the fluid physical transformations that are associated with young growing bodies. In contrast to the case of Tiresias, there is no fear within the fiction that these prophets might contaminate the minds of their listeners; rather, a nondiegetic fear is raised (by parents, teachers, and other children's advocates) concerning the minds of young television spectators, which might be corrupted by the violent behavior of these transformers and by their crass commodification of growth.

The primary difference between the myths of the Turtles and Rangers is the choice of transformational trope, mutation versus morphing. Without dismissing the distinction between these species as specious, I would reject Frye's normative contrast that elevates metamorphosis to the status of genius while ranking mutations with lowly minor talents. Whereas mutation looks backward to Darwin's nineteenth-century transformative trope of evolution, morphing looks forward to new media. Yet that does not mean that morphing is necessarily more progressive; for like Jim Carrey and like any new medium, morphing merely opens a new space onto which the old transformative myths and their power struggles can be projected and replayed.

Morphing is a high-tech mode of transformation that has more to do with empowerment than with appearance. In contrast to mutation, this mode of shape-shifting is based on technological rupture rather than being part of a "natural" process. It is active rather than passive: you do it to something or yourself rather than having it done to you. This distinction is linked to the shift in spectator position, from passive viewer (associated with movies and traditional television) to active player (demanded of video games and other interactive media, which most of these series emulate).

Thus the mutation of the Turtles is as passive as Gregor's metamorphosis into a cockroach and potentially just as grotesque. Their adventures (both in the TV series and the three movies) feature many other corporeal conversions that they themselves control—growing larger or smaller; impersonating Bogart, Cagney, and other notable stars; and going to masquerade balls where their turtle costumes are greatly admired. In fact, masquerade and voguing are fundamental to their myth, as is shown in the opening line of the original Laird and Eastman comic book, "Stupid Turtle costumes!" This parodic remark initiates a line of inquiry about the constructedness of the subject—a motif that runs throughout the myth and that is most resonant on the register of gender, particularly in the first TMNT movie.

Turtles are a species whose gender is not immediately apparent to most human observers. Unlike most superheroes, their gender specificity appears to depend totally on their costuming, weaponry, behavior, and names, which are bestowed on them by their patriarchal master, Splinter, and by the symbolic order on which his power is based. Within this myth, masculinity proves to be not biologically determined but culturally "constructed"—it is a role that can be chosen, learned, acquired, and performed even by aspiring members of so-called "inferior" species such as rats, turtles, females, kiddies, and teens. To succeed in this quest for masculine empowerment, one must undergo one or more kinds of transformation: mutation (the Turtles), martial arts training (Splinter), masquerade (April in her jumpsuit), or consumption (kids who buy into the network). Although this conception of masculinity might offer more flexibility than biological determinism, it can hardly be consoling to feminists, for the only way for a female to be empowered is to become one of the boys. Still, it implies that the performance of gender is an active form of voluntary mutation.[22]

The cultural meaning of morphing has similarly been expanded by the Power Rangers. Not only is it an electronic means of shape-shifting readily available in proliferating software programs for digital compositing that have recently made such imagery and myths seem *meta*-morphic, but it is also a successful marketing tactic for transforming a tacky Japanese sci-fi action series into the top-rated show in American children's television and a multimedia megahit in the global market. This form of cultural morphing compensates for the Japanese mastery over entertainment hardware (by companies like Sony), thereby demonstrating that the United States still has superior know-how with software! Moreover, the visual tackiness of the series (its artificial sets, corny dialogue, and crude dubbing) cultivates a precocious appreciation for kitsch and camp. Thus it implies that the more sophisticated work of meaning production is being performed by active spectators (who creatively adapt these texts to their own personal needs) rather than by crass producers (who merely recycle obsolete images from other cultures to make a quick buck). This dramatization of active spectatorship was also central to the success of *Peewee's Playhouse* and the Nickelodeon cable network, as well as to *Mystery Science Theater* and *Beavis and Butthead.* It can also be traced back to works in earlier forms and eras, such as Woody Allen's clever movie *What's Up, Tiger Lily?* (1966) and Cervantes's *Don Quixote.*

In the visual representations of the actual process of morphing, the disjunctive cutting accentuates the weapons and athletic moves of the Rangers, both of which disrupt the linear continuity of the story. Such disruptions facilitate the morphing of this action sequence into the adjacent commercials for Power Rangers hardware, which frequently use a similar visual style. These matching morphing sequences (in the series and the commercials) frequently feature a distinctive form of "power voguing" (young kids in superhero costumes striking macho poses and skillfully brandishing weapons), which can easily be replicated in bedrooms and boulevards across America (as the boy on Larchmont vividly demonstrated).

Both in the series and in the commercials, morphing functions as a form of accelerated consumerist suture. You are liberated from your infantile dependency on a "bad" mother like Rita Repulsa by learning how to become a shape-shifter who can move as fluidly as your superprotean heroes from TV images, to plastic toys, to video games, to blockbuster movies. But to fully achieve this mode of interactive spectatorship, you must first acquire the Mighty Morphin' hardware—the gloves that make karate moves with sound effects, the projectors and motion detectors that transform your own room into a battleground, and your chosen weapon that enables you to put your own personal spin on the myth.

This process of consumerist morphing is not restricted to *Power Rangers:* you can also find it in commercials for Gushers Fruit Snacks, where faces undergo morphing and humans are turned into toons as in popular hybrid movies like *Who Framed Roger Rabbit* (1988), *The Mask* (1994), and *Space Jam* (1996). It is also easily transferable to new superheroes and fads on the horizon that always threaten our reigning transformers with the kind of extinction that Kafka's Gregor so eagerly sought.

No one can accurately predict exactly how long any particular fad will last (not even a so-called marketing genius like Haim Saban), but we do know that it *will* be replaced and that the pace of the substitutions will be partially driven by children's growing consumerist desires. Like the Turtles and Rangers, all toys belong to a transformative system of marketing that empowers kids through their ever-changing choices as consumers. Like Warhol's proverbial fifteen minutes of fame or Baskin Robbins's flavor of the month, stardom in the children's toy market is merely a temporary subject position based on frequent substitutions. Occasionally there is a major innovation like Mattel's razor-blade theory of marketing, which was transferred to the toy industry to launch Barbie; in this system, an inexpensive doll functions as hardware and her proliferating accessories as faster-selling software, to create a two-tiered product with varying rhythms of replacement that keep consumers constantly coming back for more. Although Barbie is a singular superstar in the toy industry whose coding is exclusively female and nonviolent and whose shape-shifting powers are more restricted than those of the Turtles or Rangers, she provides the primary transformative model for their commercial success; for on issues of consumerism, they are merely another species of the same basic phenomenon. Yet whether we dismiss them as a minor mutation or celebrate them as a true meta-morphosis, these superheroes drive us back to mythic transformers that help us reshape ourselves as resilient players who (like Sony's Akio Morita) are ever responsive to the changing cultural moment.

Notes

1. For a perceptive essay on Transformers, see Susan Willis, "Gender as Commodity," *South Atlantic Quarterly* 86, no. 4 (fall 1987): 403–21.

2. As quoted by Mike Freeman in "Haim Saban: The 'Power' Is His," *Broadcasting and Cable* (20 December 1993): 30. Sounding like Akio Morita of Sony (in the epigraph), Saban put it this way: "We have this picture puzzle of various countries around the world, with each being able to generate a certain amount of money for certain products. And if we can make sense out of a production by mixing Korean and Luxembourgish investments that would cover the production costs, then the rest of the world is open for sales."

3. Ibid.

4. Ibid.

5. Ibid.

6. Eric Schmuckler, "Oh, What a Beautiful Morning," *Mediaweek* (18 April 1994): 31.

7. For elaborations on this connection, see Edward Rothstein, "Technology," *New York Times,* 1 April 1996, D3; and Heather Hendershot's video *Mighty Morphin' Censorship: Who's Watching Children's TV?* distributed by Paper Tiger Television.

8. Marsha Kinder, *Playing with Power in Movies, Television, and Video Games: From Muppet Babies to Teenage Mutant Ninja Turtles* (Berkeley and Los Angeles: University of California Press, 1991), 135.

9. Before turning to Greek and Roman mythology, I can't resist citing a couple of examples from Polynesia to demonstrate that the contemporary appropriation of traditional mythological transformers and the reinscription of their gender and sexuality is not restricted to the West. In her award-winning first novel, *Where We Once Belonged* (Auckland: Pasifika Press, 1996), Samoan writer Sia Figiel draws on her culture's traditional myths about the powerful god Pili the Lizard, whose "ability to metamorphosise . . . was the one and only gift given him by his father Tagaloaalagi [the creator]" (39). Like the Ninja Turtles, this reptilian transformer sustains his masculine powers through compulsive consumption, but on a cosmic (rather than a multinational) scale: "To maintain his strength, Pili ate stars, drank

oceans, ate planets . . . ate, ate, ate," (139–40). In Figiel's reinscription, her young female protagonist Alofa compares Pili with Aolele, the beautiful earth goddess after whom he lusted and whose seven brothers defeat him, transferring his transformative power to their sister. Alofa reads her own adulterous father through the myth of Pili and conversely the god through her father. By appropriating the myth, she challenges her patriarchal family and culture: "I heard the story of Aolele for the first time when I was in the womb, and because of it I willed myself female" (145).

The gender mobility of the shape-shifter trope is also central to Vilsoni Hereniko's play *Fine Dancing* (1997), which features a moon goddess Hina, who is female in act 1, male in act 2, and androgynous in act 3, and who, like Morpheus, is capable of assuming the form of the dreamer's loved ones. Performing the gender liminality of the Tongan *fakaleiti* (or Samoan *fa'afafine*), a clowning figure who is traditional in many Pacific island cultures (including Hereniko's own homeland of Rotuma), Hina generates transformations of cultural identity not only in a runaway wife and her abusive husband but also in a drag queen who helps the goddess reveal the specific Polynesian inflection of these transformative tropes.

10. *The Odyssey of Homer,* trans. Richmond Lattimore (New York: Harper and Row, 1965), 75–76.

11. Although Tiresias's gender bending figures prominently in a modernist poem such as Eliot's "The Waste Land," which describes the seer as a blind old man with withered female breasts who has suffered the sexual tribulations of both genders, in most classical texts, it is represented merely as a means of acquiring his omniscience, which is the real source of his power. T. S. Eliot, "The Waste Land" (1992), in *T. S. Eliot: The Complete Poems and Plays (1909–1950)* (New York: Harcourt Brace, 1934), 43–44.

12. Sophocles, *Oedipus the King,* trans. David Grene, in *Greek Tragedies,* vol. 1, ed. David Grene and Richmond Lattimore (Chicago and London: University of Chicago Press, 1960), 122–23.

13. For example, his own unique condition of physical blindness and singular knowledge prefigures Oedipus's fate, a connection Tiresias makes explicit in his first encounter with the tragic hero: "You are a poor wretch to taunt me with the very insults which every one soon will heap upon yourself" (126). Similarly, in *The Bacchae* (c. 405 B.C.) by Euripides, Tiresias's masquerade as a bacchanalian reveler is cruelly mocked by the blasphemous king Pentheus. Not only does the old seer's masquerade evoke his own earlier incarnation as a woman, but it also prefigures the king's fatal cross-dressing, which leads to his brutal death at the hands of his own deluded mother and her band of enchanted bacchae, who see

through his drag but without recognizing his royal identity. In contrast to the mercurial masquerades that the bodies of the prophets and gods undergo, the tragic transformations of these royal figures are irreversible.

14. Gilles Deleuze, *Masochism: An Interpretation of Coldness and Cruelty,* trans. Jean McNeil (New York: George Braziller, 1971).

15. "The lunatic, the lover, and the poet / Are of imagination all compact. / One sees more devils than vast hell can hold: / That is the madman. The lover, all as frantic, / Sees Helen's beauty in a brow of Egypt. / The poet's eye, in a fine frenzy rolling / Doth glance from heaven to earth, from earth to heaven; / And as imagination bodies forth / The forms of things unknown, the poet's pen / Turns them to shapes, and gives to airy nothing / A local habitation and a name." William Shakespeare, *A Midsummer Night's Dream,* 5.1.11.7–17. Ironically, the very capacity for transformative imagination that enabled both Othello and Desdemona to love each other across great barriers of age, race, culture, and experience also makes him susceptible to Iago's corrosive conceits, which (unlike the playful spells in *A Midsummer Night's Dream*) cannot be erased from the mind of the tragic hero. Iago's power as a villain lies precisely in his ability to poison the imagination of this proud ruler (as Tiresias did to Oedipus and Creon, but in this case with malicious intent). His conceits generate painful transformations—of Desdemona into a whore and Othello into her insane killer (just as Tiresias's words changed Oedipus's beloved wife into his incestuous mother and himself into a patricide, and Pentheus's mother into his killer).

16. William S. Burroughs, *Naked Lunch* (New York: Grove Press, 1959), xxxvii.

17. Ibid., xxxix.

18. Ibid., 54–55.

19. Quoted by Schmuckler, 28.

20. Ishmael Reed, *Mumbo Jumbo* (New York: Avon Books, 1972), 12.

21. James Kaplan, "More Power," *TV Guide* (29 October–4 November 1994): 38.

22. For an elaboration of this issue, see chapter 4 in Kinder, *Playing with Power.*

II. Transformation, Technology, and Narrative

5

A Brief History of Morphing

MARK J. P. WOLF

Over the past few years, "morphing" has become a catchphrase denoting a particular state-of-the-art graphic special effect and is among those elements defining the visual style of the 1990s. Morphing creates fluid and often surreal transformations by combining the cross-dissolving and the warping of images (in which certain areas are stretched or compressed). During a morph, an object seems to reshape and transform itself gradually into another object in full view of the audience, providing the same kind of pleasure one might find in a well-performed trick of stage magic. Morphing can be found in a variety of media, including CD-ROMs like *Myst* (1993); films including *Terminator 2: Judgment Day* (1991), *Star Trek VI: The Undiscovered Country* (1991), *The Fifth Element* (1997), and *Men in Black* (1997); and on television in programs such as the series *Star Trek: Deep Space Nine* and *Roar,* in music videos like Michael Jackson's *Black or White* (1991), and in an ever-increasing number of TV commercials. Although nearly all of these media use morphing for transformations of characters and objects, morphing also has formal possibilities as a transitional device, akin to dissolves and match cutting, and other new uses and techniques that have yet to be fully explored. Ideas concerning visual metamorphoses have themselves grown and changed over the years and today seem to have found their fruition in the "morphing" of the computer age, where they have begun to flourish.

Origins of Morphing in Art and Science

The extent to which morphing is used today and the degree of graphic detail now possible are both due to more recent technological advances, but some of the ideas behind morphing and their applications have been around for several centuries. Leonardo da Vinci and other artists of the Renaissance studied human proportion, often doing their own dissections and measurements, and noticing correspondences and ratios in the human form. Generalizing the proportions of the human body led to the diagramming of anatomical ratios, and a more mathematical approach to the drawing of the human form.

During the early 1500s, Albrecht Dürer also applied lines of proportion to the human head and face and found that by repositioning these lines at various distances or even rotating and moving their angles relative to one another, he could account for a wide variety of facial structures. One page of his notebook sketches contains a series of heads in profile that are all variations derived from an initial set of proportions, represented by horizontal and vertical lines. In each successive drawing, the spacings of the grid lines are altered, and the features are drawn accordingly. Anticipating feature-based morphing (in which a subject's features can each be morphed individually), certain lines, like those indicating the angle of the nose, are altered independently of the horizontal and vertical grid lines. Through these sketches, Dürer revealed the underlying correspondences between facial structures, and in his *Four Books on Human Proportion,* he sketched out both the idea and method of transforming one head into another based on mathematical ratios.[1] The notion of correspondences between different forms became the central idea behind morphing, but a way of precisely expressing such transformations in mathematical terms, however, would not come until after Dürer's death.

In 1637 René Descartes's *Geometria* appeared, which combined the fields of algebra and geometry and laid the foundations of analytic geometry.[2] The book also introduced the notion of the grid and the system of Cartesian coordinates, which could locate points and lines on the grid. Points could be expressed as the intersection of two coordinate lines, x (horizontal) and y (vertical), and lines were represented as equations (such as $Ax + By = C$). Changes in position and size could also be expressed mathematically. The Cartesian grid mathematized space and allowed for a precise expression of the manipulations of geometric forms within it. Through its use, the deformations with which Dürer had experimented

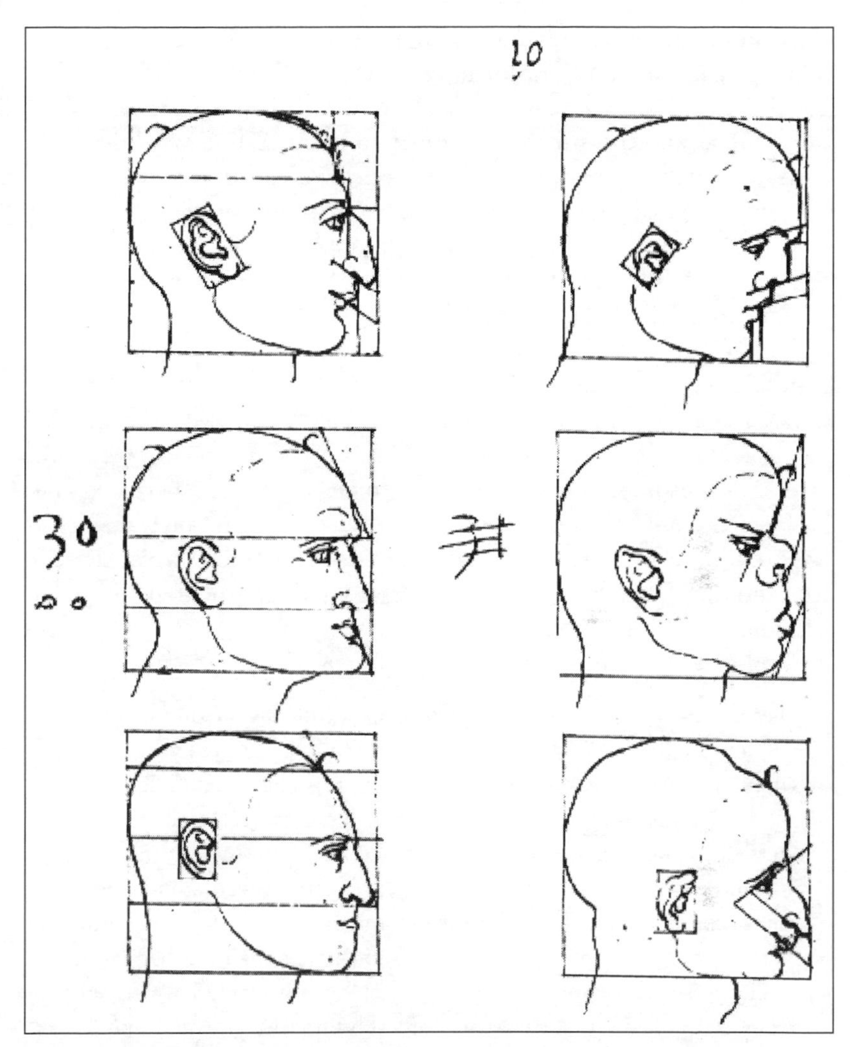

Figure 5.1 By altering the spacings of grid lines representing the proportions of the human head, Albrecht Dürer found a way to express the differences between facial structures. His use of diagonal lines that could be moved independently of other grid lines anticipated feature-based morphing. From *The Human Figure by Albrecht Dürer: The Complete Dresden Sketchbook.* Courtesy of Dover Publications, Inc.

could be quantitatively compared to each other and with the original source image undergoing distortion.

The Cartesian grid allowed mathematicians to visualize mathematical formulas and transform one form into another, but for many years, these transformations were mainly concerned with abstract figures, points, lines, and curves. It would be some time before more complex forms, such as those Dürer was working with, would again be applied to flexible grids. Today the idea of a grid and coordinate system is the basis of computer graphics and digital imaging.

The mathematics of analytic geometry were finally combined with Dürer's studies of human physiognomy in D'arcy Wentworth Thompson's seminal work in biological morphology, *On Growth and Form*, first published in 1917, with a second revised edition appearing in 1942. In the book, Thompson even makes reference to Dürer's images while explaining his process. The book's final chapter, "On the Theory of Transformations, or the Comparison of Related Forms," examines similarities and correspondences between natural forms and in their growth patterns. His description of his method indicates that it was amazingly close to current methods used for computer-based morphing:

> Let us describe in a system of Cartesian coordinates the outline of an organism, however complicated, or a part thereof: such as a fish, a crab, or a mammalian skull. We may now treat this complicated figure, in general terms, as a function of x, y. If we submit our rectangular system to deformation on simple and recognized lines, altering, for instance, the direction of the axes, the ratio of x/y, or substituting for x and y some more complicated expressions, then we obtain a new system of coordinates, whose deformation from the original type the inscribed figure will precisely follow. In other words, we obtain a new figure which represents the old figure under a more or less homogeneous *strain*, and is a function of the new coordinates in precisely the same way as the old figure was of the original coordinates x and y.[3]

Among the deformations Thompson goes on to discuss are stretching and compressing along the *x* or *y* axes, tapering, shearing, and conversion to radial coordinates (wherein one set of coordinates lies around an arc or circle instead of a straight line). Throughout the chapter, he provides numerous illustrations of biological deformations, including fish, crabs, bones, leaves, and the skulls of mammals. One illustration, showing the structures of the pelvis of archaeopteryx and of apatornis, even goes so far as to include an artist's speculatory drawings of three "transitional types" interpolated between the two bones, to demonstrate a possible transition from one form to the other. Had these grid-based images been shot as an animated sequence, they would have constituted a morphing sequence not unlike those done on the computer today.

Toward the end of the chapter, Thompson speculates about taking the process into the third dimension:

> In this brief account of coordinate transformations and of their morphological utility I have dealt with plane coordinates only, and have made no mention of the less elementary subject of coordinates in three-dimensional space. In theory there is no difficulty whatsoever in such an extension of our method; it is just as easy to refer the form of our fish or our skull to the rectangular coordinates x, y, z, or to the polar coordinates ξ, η, ζ, as it is to refer their plane projections to the two axes to which our investigation has been confined. And that it would be advantageous goes without saying, for it is the shape of the solid object, not that of the mere drawing of the object, that we would want to understand; and already we have found some of our easy problems in solid geometry leading us (as in the form of the bivalve and even of the univalve shell) quickly in the direction of coordinate analysis and the theory of conformal transformations. But this extended theme I have not attempted to pursue, and it must be left to other times, and to other hands.[4]

The other times and hands turned out to be those of computer animation. Interestingly enough, Dürer also seemed to have had considered taking his explorations into three dimensions, as evidenced in his drawings. On one of the pages in his sketchbook, alongside his grid-based profiles, there are drawings of two heads made up of polygons, which are eerie premonitions of three-dimensional polygonal-based computer imaging.

Thompson's seminal work not only established the usefulness of morphing in biology but also showed how transformations could be accomplished using the Cartesian coordinate grid that allowed the quantification of differences between various deformations. Furthermore, both Thompson's and Dürer's work show that the ideas for three-dimensional morphing were around for several centuries before the technology was available to do all the necessary calculations or to set the images of transformation in motion.

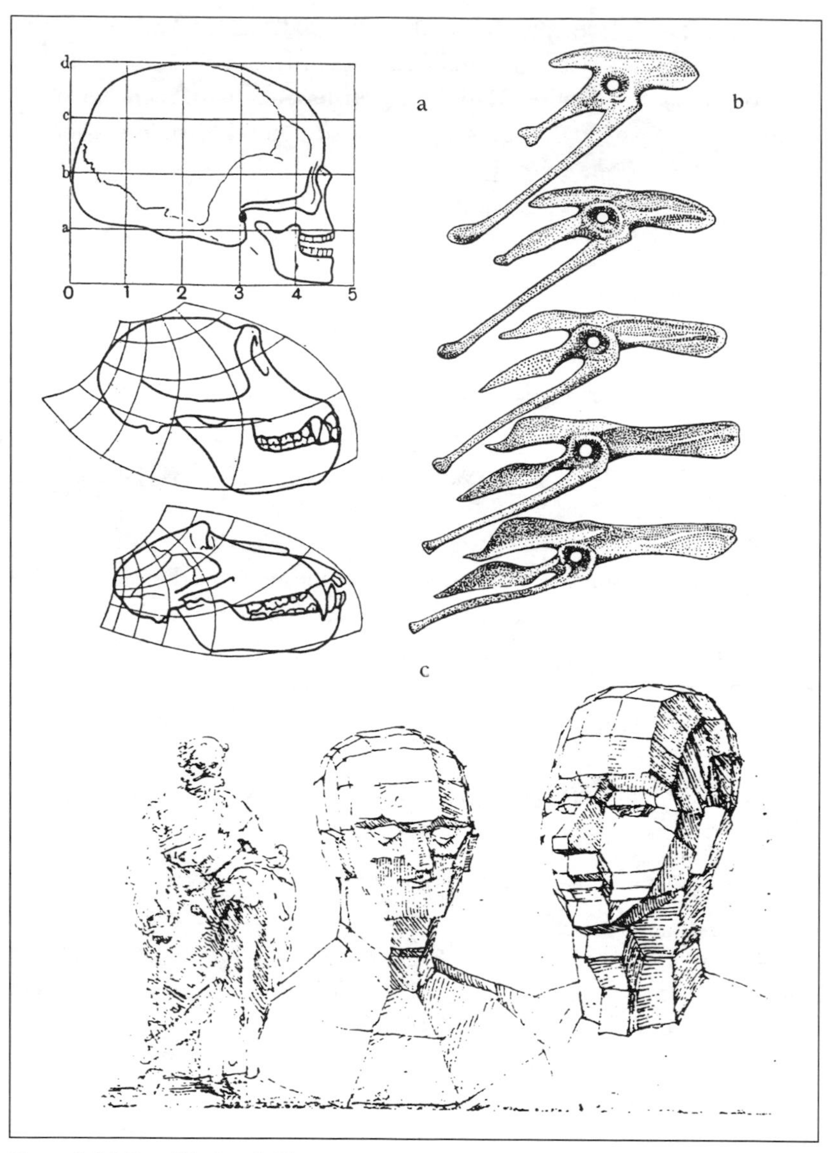

Figure 5.2 D'Arcy Wentworth Thompson used mathematical grid deformation to examine structural similarities between different species. In *a,* grid lines show topological similarity between skulls of a human, chimpanzee, and baboon; in *b,* Thompson interpolates bone shapes between the pelvic bone of the archaeopteryx and apatomix, anticipating the interpolated images between the end point images of a morph. In *c,* almost five hundred years before polygonal-based computergraphic imaging, Albrecht Dürer sketched heads made of polygons, suggesting facial deformation in three dimensions. From *On Growth and Form,* by D'Arcy Wentworth Thompson. Courtesy of Cambridge University Press.

Morphing in the Cinema

The ability to give images the illusion of motion came, of course, with the cinema. Whereas the concepts discussed in the previous section were all limited to the purposes of art and science, other nonmathematical kinds of morphing appeared in the cinema. Hand-drawn animation, as early as Emile Cohl's *Fantasmagorie* in 1906, often explored fluid changes and elastic transformations; and later other techniques using sand, clay, and other media also emphasized these abilities. In live-action cinema, make-up effects combined with editorial devices were used for character metamorphoses, like Lon Chaney Jr.'s transformation from human to werewolf in *The Wolf Man* (1941). For live-action morphing in these films, either cross-dissolves or clever cutting was used (characters would pass in and out of shadow or behind a series of pillars or trees as they changed), resulting in an edited transformation as opposed to an unbroken sequence of changes occurring in full view. More recent films, such as John Landis's *An American Werewolf in London* (1981) and John Carpenter's remake of *The Thing* (1982), used mechanical effects to show character transformations occurring within a single shot, such as heads stretching or body parts lengthening. In addition to this, *Starman* (1984) featured a sequence in which the alien starman metamorphoses from an infant to a full-grown adult in a matter of seconds; in a single shot, the starman's head stretches vertically and then widens horizontally into its adult form. The effect was produced through the use of *replacement animation*, a process in which a series of three-dimensional models, each slightly different from the next, are photographed one after another to produce an illusion of motion, in much the same way a series of drawings is photographed in conventional animation. For the metamorphosis shot in *Starman*, a series of life-size molds representing incremental stages between the child and adult starman were positioned on the set and photographed one after the other to bridge the gap between the two actors playing the starman as infant and adult. All these transformations, however, were done in a series of close-ups and cuts, allowing for different parts of a transformation to be completed separately, with no single shot containing more than a step in the process. Considering what technology was available, films such as these took mechanical effects as far as they could go, and the next advances to take place would use a computer.

The potential for the use of computer graphics for visual morphing was recognized early in the development of computer technology. Painter Peter Földes used computer imaging to transform line drawings into each other in his short films *Visages des femmes* (1968), *Metadata* (1972), and *Hunger* (1973). Földes would create line drawings of key frames and then use the computer to interpolate a sequence of images between them, resulting in transformations in which the lines of the images would shift, grow, or shrink as one image became the next. His films explore both the visual and metaphorical potential of these transformations. Földes spoke of his particular interest in metamorphosis and the computer in a communication to Giannalberto Bendazzi:

> The art of the twentieth century is cinema. The language of the twentieth century is technology. In my films, I made metamorphosis. *Visages des femmes* was a perpetual metamorphosis, created by handmade drawings. With a computer, I can still make metamorphoses, but with greater control over each line of the drawing, which I can move as I please. And I work faster, because the machine frees the artist from the fatigue of labor. A miniaturist can work for seven years on a single work; nobody says that Rembrandt's paintings are less beautiful only because he spent less time on them.[5]

During transformations in Földes's films, lines move independently of one another, unlike the continuous deformations of the grid found in Thompson's transformations, in which lines retained their relative positions as forms underwent change. In Thompson's transformations, objects being compared had inherent similarities; skulls were transformed into other skulls, fish into other fish, and so forth, so there was little need actively to *invent* correspondences between the two objects being compared. In Földes's films, however, one object could metamorphose into any other object, regardless of how different they initially appeared from each other. In this sense, Földes's work anticipated *feature-based morphing* in which correspondences are chosen by the artist, allowing objects without any inherent similarities to be morphed from one to the other.

Other morphing research went on in the early 1980s, such as the work of Tom Brigham at MIT, which involved the mathematizing of the process and extending it beyond two dimensions. The basics of three-dimensional morphing were being explored on the computer; polyhedra such as cubes, spheres, pyramids, and other geometric primitives could be warped and changed into each other and represented an attempt to bring coordinate-based morphing into the third dimension. The objects used in these metamorphoses, however, were all created in the computer and were geometrically simple and featureless, and the transformations were likewise straightforward interpolations. These kinds of transformations were used not only in computer-animated short films of the period but even in the feature film *Tron* (1982), where the character Bit changes shape while saying "Yes" and "No."

Although there was limited use for the morphing of simple polyhedra, the verisimilitude of live-action cinema required photo-realistic effects that computer systems were not ready to offer until the late 1980s. During the 1980s, there were a few exceptional appearances of photo-realistic computer graphics in feature-length live-action filmmaking; the "glass man" effect in *Young Sherlock Holmes* (1985), in which a stained-glass knight comes out of a window and walks around, and the "water arm" or pseudopod effect in *The Abyss* (1989), in which a long, translucent arm of water confronts and interacts with the film's human characters. Apart from these exceptions, however, most computer-generated objects used in film were still noticeably computer graphics (geometric, polygon-based or vector-based, flat colors lacking texture, and so on), and often appeared diegetically as such: for example, the computer displays in *Return of the Jedi* (1983) or *Wargames* (1983), or the video game–like vehicles and spaceships in *Tron* (1982) and *The Last Starfighter* (1984). For more photo-realistic effects in film to be possible, advances in computer hardware and software were needed, as well as film technology that would be able to seamlessly combine computer-generated effects with the live-action worlds seen on-screen.

Digital Film and the Morphing Revolution

With special effects by Industrial Light and Magic (ILM), *Willow* (1988) became the first film to feature photo-realistic morphing effects. Using "Morf," a proprietary computer program developed by ILM graphics artist Doug Smythe, *Willow* featured a series of morphs of a character changing first from one type of animal to another, and finally to a human being. The transformation was the most complex photo-realistic metamorphosis ever to appear on-screen in a live-action film; although the sequence is broken up into two main shots, each of them features multiple transformations. The creation of the sequence involved a combination of computer graphics with physically based methods including models and puppetry and demonstrated that photo-realistic morphing effects done on a computer were finally possible in the cinema. Animatronic puppets were used for the creature at various stages, and shots of them were digitized into a computer, where morphing software was used to create the transitions between them. A similar technique was used a year later during the "Donovan's Destruction" sequence in *Indiana Jones and the Last Crusade* (1989), in which a character withers away into a skeleton.[6]

Industrial Light and Magic has been in the vanguard of special effects technology since its inception (it was first organized to do effects for *Star Wars* [1977]), but it was only a short while after *Willow* appeared that morphing techniques became more widespread throughout the effects industry. Two technologies made this possible: the appearance of digital film systems, and the development of feature-based image metamorphosis software. Pacific Data Images (PDI), a computer animation company, also became a leader in the morphing revolution in film. In 1990 PDI was the first computer animation studio to commercialize feature-based morphing for advertising and music videos, and the first to offer digital film scanning capabilities in the Los Angeles area.

The following year, *Terminator 2: Judgment Day* (1991) appeared, with its "chrome man" effects for the T-1000 terminator character, who morphed through various shapes and into different people with quicksilver fluidity. Both ILM and PDI worked on the film, and ILM won an Academy Award for their work. *Terminator 2* was also a very high-profile film and helped to bring morphing to the attention of a widespread audience (*Willow* had good effects, but the film turned out to be a flop). Unlike the morphing sequence in *Willow,* however, the morphing in *Terminator 2* was not feature based (features disappeared and reappeared during morphs, instead of changing directly from one to another); although the end points of his morphs were photo-realistically colored objects or people in color, the T-1000 character always turned to featureless chrome during his transformations. Although the effect was used many times throughout the film, it occurred in a wide variety of conditions and situations, and not just in an isolated sequence.

In 1992 advances in computer hardware and software brought morphing to the film industry at large. A commercial digital film system, the Kodak High Resolution Electronic Intermediate System (also known as the Cineon), became available. With the system, a filmmaker can scan film into a computer, alter the images digitally, and then print the images back out onto film with no visible loss of image quality. This lack of generational loss is possible due to high-resolution imaging; at full resolution, the Cineon can scan and print film at 166.67 pixels per millimeter.

Since 1992 a number of companies besides Kodak now scan and print film, and digital effects have quickly become quite common. Through digital imaging, the picture becomes a grid of pixels that can be deformed and warped mathematically as were the gridded images used in Thompson's *On Growth and Form.* Rather than making metamorphoses occur physically or drawing them on paper by hand, they can now be calculated, recalculated, and experimented with on a computer before reaching their final form. Once the hardware existed to produce a mathematically manipulatable image, all that remained to be added was the software to manipulate it. Although ILM had developed a morphing program, it was proprietary and not available outside the company. In 1992 Thaddeus Beier of Silicon Graphics Computer Systems and Shawn Neely of PDI published a paper entitled "Feature-Based Image Metamorphosis," which described the morphing methods used by PDI and allowed other companies to develop their own. (PDI had been using the software for two years and had already had great success with it; at the end of the paper, the authors thank the owners of PDI "for allowing us to publish the details of a very profitable algorithm.")[7]

The Morphing Process

There are two basic steps in the morphing process: warping the images into matching shapes, and then cross-dissolving them. Dissolving by itself is simple; the colors of the pixels of one image are shifted to those of the pixels in the second image. Because the colors of the pixels may be very different, a great deal of color shifting must often take place. If the images are warped so that features of the first image are dissolved to features of similar shape and color in the second image, the whole transition is much smoother and less disrupting or jarring to the image. The difficult part of the morph is the warping of the images so as to increase the correspondence between their features; regular deformations of the grid and mesh warping such as Thompson used will not work very well, if at all, when the source and destination images of the morph are too disparate. Feature-based morphing allows the user to choose direct correspondences and match features between images regardless of their differences.

The process begins with the user's designation of pairs of line segments that mark the user-selected corresponding features in the two end point images (correspondences can be based on the size, color, shape, position, function, and so on, of features within the images). One line segment is placed on the source image, over a particular feature, such as an eye, the bridge of a nose, the side of a head or chin, and so on. The second line segment is placed on the corresponding feature within the destination image (into which the first image is to be morphed). Pairs of line segments like these are placed on the image until there are enough to define and delineate the transformation adequately, and they can later be readjusted if necessary. Next, the pairs of segments in the two images are interpolated to find the intermediate line segments, which are halfway between those in the first and second images. This set of line segments will form an intermediate grid at the halfway point of the morph. Each of the end point images is then warped to fit the intermediate grid. Each of the end point images is deformed until its line segments coincide with the intermediate grid's line segments. As each line moves, the grid area around it is also affected; the closer a pixel is to the line, the more it will be affected; while pixels farther away are less affected. The overall effect is similar to the field lines of influence produced by a magnet, and so this process is sometimes referred to as "magnetic sculpting." The two end point images that are warped to match the line segments of the intermediate grid are then cross-dissolved throughout the duration of the morph. The warping and cross-dissolving occur simultaneously, until the morph has completed its transition to the second image.

The process described is that of a morph between two still images, but the method can be extended to moving imagery as well. Toward the end of their paper, Beier and Neely write:

> It is often useful to morph between two sequences of live action, rather than just two still images. The morph technique can easily be extended to apply to this problem. Instead of just marking corresponding features in the two images, there needs to be a set of line segments at key frames for each sequence of images. These sets of segments are interpolated to get the two sets for a particular frame, and then the above two-image metamorphosis is performed on the two frames, one from each strip of live action. This creates much more work for the animator, because instead of marking features in just two images he will need to mark features in many key frames in two sequences of live action. For example, in a transition between two moving faces, the animator might have to draw a line down the nose in each of 10 key frames in both sequences, requiring 20 individual line segments. However, the increase in realism of metamorphosis of live action compared to still images is dramatic, and worth the effort. The sequences in the Michael Jackson video, *Black or White,* were done this way.[8]

PDI did the effects in *Black or White*, which features thirteen different people in close-up, whose heads smoothly morph from one to the next as they sing the chorus to the song. The people range in age, race, gender, and facial structure, and the resulting morphs of them are quite dramatic. This ability to morph one sequence of moving imagery into another was a major breakthrough, the potential of which has yet to be fully explored in cinema.

Beier and Neely's paper not only opened morphing up to a wider audience but also encouraged further research into morphing methods. One paper from 1995, "Feature-Based Volume Metamorphosis," by Apostolous Lerios, Chase D. Garfinkle, and Marc Levoy, extends Beier and Neely's technique into the third dimension by describing a method for morphing between 3-D computer-generated models, rather than just between 2-D images of those models or images scanned into a computer. Additional difficulties are present in three dimensions; in addition to creating a smooth transition from one object to the next, Lerios, Garfinkle, and Levoy write that the morphs "should be realistic objects with plausible 3-D geometry . . . which retain the essential features of the source and target."[9] Their technique is similar in concept to Beier and Neely's, but with some additions. Instead of using only line pairs to designate features, they use points, line segments, rectangles, and boxes to define the elements or features used for the morph. Their magnetic sculpting takes place in three dimensions, where features can rotate and change orientation as they move around during a morph. Their method provides more control over the cross-dissolving process than does Beier and Neely's algorithm, allowing the dissolve to speed up or slow down as the user desires. The paper also includes a description of the user interface, and an algorithm that is said to help speed up performance of the process by a factor of fifty.

Powered in part by its immense value to the entertainment industry, morphing research has grown a great deal in recent years, and its fruits are beginning to appear on-screen. These and other advances are certain to have their impact on cinematic special effects in the coming years, just as advances in mechanical and optical effects did. Digital morphing effects, however, are not without their limitations.

That's a Stretch! The Limitations of Morphing

A good morphing sequence is a difficult thing to achieve, and techniques used to produce it must be taken into consideration during the shooting of the live-action footage needed for the sequence. For most films using computer-animated effects, it is common to have a technical director from a computer effects company present on the set during the shoot to make sure that the director gets everything needed for the effects to be done properly, and also to suggest what may or may not work regarding the computer effects. The position, intensity, specularity, and other data for each of the light sources in the shot is often recorded, as well as measurements of the set. Sometimes footage of ball-and-stick cubelike structures that act as a standard for measuring perspective are shot on the set, and the images of them are later used as a guide when computer-generated models are added into the shot. But even if one ignores the extra expense during production and postproduction of doing state-of-the-art special effects, morphing effects present other limitations for the filmmaker.

Image metamorphoses operate on flat two-dimensional images rather than on three-dimensional representations of objects or scenes, and owing to this, the perceived dimensionality of the image can be affected. In other words, the interpolated images between the end point images may not always retain a visually coherent three-dimensional structure, and depth cues can be momentarily lost or inconsistent during the morphing sequence. Another problem is the separation of the foreground object, which is undergoing the morph, from the background in which that object appears. To perform a morph on an image of a character or object without affecting the background behind it, a separation between foreground and background elements must take place so that only the foreground element is affected by the morph. Likewise, during the morph, unseen parts of the background behind a character—areas that were not recorded by the camera because they were hidden behind the actor during the entire shot—might be revealed. Because the image is only a grid of pixels, the image provides no information about what objects were hidden behind others at the time the image was photographed, and so the background must be shot separately to fill in behind the foreground image when necessary. This can be accomplished in two ways, depending on what is needed in the scene.

In one method, the foreground element is shot separately in front of a blue screen (a solid color screen against which the foreground object is filmed) and later composited onto a background plate that is shot separately. This can be done either optically or digitally and is the process used for most effects involving the addition to the shot of characters or objects that were not present when the scene was filmed: for example, the cartoon characters in *Who Framed Roger Rabbit?* (1988), or the snow speeders in *The Empire Strikes Back* (1980).

Another method, made possible by digital technology and the motion-control camera, was used in Robert Zemeckis's *Death Becomes Her* (1992). The motion-control camera, first used in *Star Wars* (1977), is a boom-mounted camera on a dolly resting on tracks and connected to a computer. Every movement the camera makes is recorded by the computer, and through the use of servomotors, camera moves can be mechanically repeated with frame-by-frame exactness. In *Death Becomes Her,* one character gets shot at and afterward has a large hole through the center of her body through which the background of the scene is visible. To achieve this kind of effect, the actor is first filmed on the set performing the scene, and the camera moves are carefully recorded by computer. The scene is shot again without the actor, the camera moves replayed exactly, and a background plate is made. On the computer, this empty background is then placed behind the shot with the actor in it, and when parts of the actor are erased, the empty background shows through and completes the image. When a larger object morphs into a smaller one, or into something with a different outline, an unrecorded area of the background hidden behind the foreground element is also revealed, so that this method can also be used when the person being morphed appears in the scene as opposed to being bluescreened into it. Morphing, then, often suffers from many of the limitations that composite imagery in general does, because it requires a separation of the foreground and background elements. Some of these limitations include restricted interaction with the environment, difficulty in matching foreground elements to the backgrounds, and the need for careful planning during production.[10]

Other limitations reside in the morph's emphasis on surface; the image is manipulated flatly, and a sense of depth can be lost in the process. Lerios, Garfinkle, and Levoy compare 2-D and 3-D methods and point out some of the shortcomings of image morphing:

> In 3D morphing, creating the morphs is independent of the viewing and lighting parameters. Hence, we can create a morph sequence once, and then experiment with various camera angles and lighting conditions during rendering. In 2D morphing, a new morph must be recomputed every time we wish to alter our viewpoint or the illumination of the 3D model. 2D techniques, lacking information on the model's spatial configuration, are unable to correctly handle changes in illumination and

> visibility. Two examples of this type of artifact are: (i) Shadows and highlights fail to match shape changes occurring in the morph. (ii) When a feature of the 3D object is not visible in the original 2D image, this feature cannot be made to appear during the morph; for example, when a singing actor needs to open her mouth during a morph, pulling her lips apart thickens the lips instead of revealing her teeth.[11]

Interestingly, here is one advantage of the mechanical methods used in the transformation scenes of *Starman* and *An American Werewolf in London;* things shot on the set need no added efforts to make them appear realistically three-dimensional.

As Lerios and the others point out, feature-based volume metamorphosis does solve a number of the problems found in 2-D image morphing, allowing the morphing objects to be treated like three-dimensional objects as regards lighting, shadows, reflections, and viewing angles, as well as interior detail like the singer's teeth. But volume-based metamorphosis techniques have their technical limitations as well, often in addition to being more expensive and time-consuming than 2-D image-based techniques. Whereas flat images are manipulated quickly and relatively easily, much more memory and computing power is needed to work with volume data. There is also the question of how the data is acquired; flat images are easily scanned into a computer, but the inputting of volume data is more complex.

In their paper, Lerios, Garfinkle, and Levoy list four methods of inputting volume data: scanned volumes, scan-converted geometric models, interactive sculpting, and procedural definition. Scanned volumes are the product of different scanning technologies, such as computerized tomography (CT), and magnetic resonance imaging (MRI). Although these technologies are fine for obtaining structural data such as the shapes of skulls (one of the examples the authors use in the paper), scanned volumes do not reveal anything about the visual appearance of the object, just as an x ray does not reveal the color or texture of a patient's skin (an ironic reversal of the usual surface-at-the-expense-of-substance problem). Scan-converted geometric models are useful but require the scanned objects to be easily represented geometrically. Interactive sculpting (creating a model though the use of a drawing or sculpting interface) and procedural definition (defining of a model though typed commands and mathematical description) both involve creating models directly in the computer as opposed to scanning data into it. Both methods rely on the ability of an artist or programmer to create (or recreate) the desired object or character, which may be too difficult or time-consuming to reproduce adequately. For example, the shapes and likenesses of human beings are sometimes scanned in and appear as computer-generated objects, such as Robert Patrick in *Terminator 2* or Denzel Washington in *Virtuosity* (1995), but the effect is often very noticeably a computer graphic and too lacking in subtlety to be a convincing replacement for the actor. The inputting of volumetric data remains an important step in the 3-D morphing process, and no method is without its shortcomings.

There are other limitations of image morphing that volume morphing does not escape; it also usually involves a composited image, it can be quite expensive due to the great amount of computer work involved, and it, too, requires a great deal of careful planning on the set. And even if all the technical problems with image morphing or volume morphing were solved, other limitations would still remain, perhaps most noticeably in terms of the narrative context in which the morph appears.

With the exception of morphs that are used as abstractions or as transitional devices, most morphs involve characters (or objects) that are within the diegetic world of the film or television program in which they appear. The often unreal nature of the morph sets a certain fantastic tone, and without sufficient explanation for its appearance, it can erode the plausibility of a scene and disrupt its verisimilitude. Despite this—or rather, because of it—the morph is often presented as the highlight of a scene, the magic trick performed right before the audience's eyes. In most of the films and television shows that use it, morphing is foregrounded not only spatially but temporally; that is, whenever a morph occurs, quite often all the narrative action around it stops, and the morphing effect momentarily takes center stage, before the action resumes. Morphs tend to occur quickly, moving from one end point image to another, and the points where they end and begin are usually made distinct and very noticeable. Perhaps most noticeable are morphs that are not feature based; in *Terminator 2*, the T-1000 loses his color and detail during morphs, reverting to a chrome quicksilver, amorphous in form. Likewise, Odo on *Star Trek: Deep Space Nine* loses all detail and features during his morphs, becoming a brown liquid. These amorphous morphs seem to follow fewer rules than feature-based morphs, which attempt to maintain some continuity between their end point images. Used in this manner, morphs can become brief interludes of special effects prowess, disrupting the narratives in which they appear. Sometimes this is done intentionally, of course, to emphasize the spectacular nature of special effects used, but it may call attention to the filmmaking at the expense of the film's diegetic world.

The fantastic nature of the morph and the need to rationalize its occurrence in the narrative have until now limited the genres in which it is found to animation (Felix the cat cartoons and many Fleischer films), science fiction *(Terminator 2, Starman, Star Trek VI: The Undiscovered Country)*, horror (*The Thing* [1982]), and fantasy (*Willow* and *Heavenly Creatures* [1994]). Even within these genres, morphing is usually used for character transformations, which are often motivated as a means of disguise or deception (shape-shifters hiding from their enemies, or disguising themselves for a surprise attack), or for utility (Odo or the T-1000 Terminator using their shape-shifting to do things they wouldn't be able to do otherwise). Because morphing effects are (still) expensive, few are allowed to be inconspicuous, although this may change as they become more commonplace and less expensive. But the need to explain the overt appearance of morphing effects in the narrative will still remain regardless of how well the technology can perform them.

The morph does have nondiegetic uses as a transitional device, alongside the dissolve and the match cut. But unlike these editorial devices, the morph affects the image in such a way that we are reminded of its nature as an image, an elastic grid of pixels, and so its usefulness as a transitional device may also be somewhat limited, although it has not yet become an accepted convention. With a more widespread understanding of the morphing process, audiences are likely to expect even higher standards from film special effects and no doubt will get them. Although home computer programs will inevitably lag behind state-of-the-art software, they can make people think about the plasticity of the image and may make audiences more critically aware of image manipulations and possible deceptions.

As morphing's spotlight is stolen by newer technologies and effects, it may become more common and less conspicuous, taking on new roles and perhaps creating new conventions in its use. As more morphing is done, different styles of morphing may well develop, increasing the expressivity of the morph, which has yet to be fully explored. Morphing may also aid in other areas such as film restoration, where missing frames or other information in a sequence of images could be approximated and interpolated.

Morphing is still a new enough technology in cinema that the novelty has yet to wear off completely. When it finally does, however, morphing may well be able to reach its full potential, both as a tool of image manipulation technology and as an element of artistic expression in the visual arts.

Notes

1. Albrecht Dürer, *Hiernn sind begriffen vier Bucher von menschlichen Proportion* (Four books on human proportion) (Nuremberg: Gedruct durch 1, 1528). Sketches from these books can also be found in *The Human Figure by Albrecht Dürer: The Complete Dresden Sketchbook,* ed. Walter Strauss (New York: Dover Publications, 1972), 231–37.

2. René Descartes, *Geometria* (1637; reprint, Leyden: Lugdon, Batavorium, Ex-Officiana Ioannis Maire, 1649).

3. D'arcy Wentworth Thompson, *On Growth and Form,* vol. 2 (London: Cambridge University Press, 1959), 1033.

4. Ibid., 1087.

5. Quoted in Giannalberto Bendazzi, *Cartoon: One Hundred Years of Cinema Animation* (Bloomington and Indianapolis: Indiana University Press, 1994), 433.

6. For descriptions of the effects used in *Willow, Indiana Jones and the Last Crusade,* and *The Abyss,* see Mark Cotta Vaz and Patricia Rose Duignan, *Industrial Light and Magic: Into the Digital Realm* (New York: Ballantine Books, 1996).

7. Thaddeus Beier and Shawn Neely, "Feature-Based Image Metamorphosis," *Computergraphics* 26, no. 2 (July 1992): 39.

8. Ibid.

9. Apostolous Lerios, Chase D. Garfinkle, and Marc Levoy, "Feature-Based Volume Metamorphosis," in *Computer Graphics: Proceedings: Annual Conference Series 1995: SIGGRAPH 95,* ed. Robert Cook (New York: ACM SIGGRAPH, 1995), 450.

10. For further details regarding the limitations of composite imagery, see my essay "In the Frame of *Roger Rabbit:* Visual Compositing in Film," *The Velvet Light Trap* 36 (fall 1995): 45–59.

11. Lerios, Garfinkle, and Levoy, 449.

6

Tracing the Tesseract: A Conceptual Prehistory of the Morph

KEVIN FISHER

Within the last decade, "the morph" has arisen as a unique phenomenological object within visual culture and, just as quickly, induced a sort of sensory fatigue through its overuse. Consequently the figural transformation of one object into another has been reduced to a sort of spectacular digital cliché. Yet on another level, the morph has moved from foregrounded effect to a subtle, often invisible manipulation of the photorealist image. It is the first, more spectacular type of transformation that this paper takes as its primary object, if only because it hyperbolizes, in a demonstrative manner, the technological and phenomenological novelty of digital morphing. Although the fluid and seamless transformations of morphing are, as I will argue, intricately tied to the specific computational properties of the digital computer, the morph's essential structure has a considerable scientific and aesthetic prehistory that may be traced to the nineteenth-century development of higher-dimensional and non-Euclidean geometries. Attempts to mathematically notate and visually represent a fourth dimension have produced various hypothetical objects, or metaobjects, of which this paper takes the *tesseract* as special case. I will spotlight specific individuals and ideas in the progressive historical extrusion of the tesseract from mathematical postulate, to object of representation, to the structural-operating principle of transformational techniques within both analog and digital cinema. To elaborate the significance of the tesseract in relation to moving images, I will examine precedents of morphing in analog cinema and seek to categorically differentiate these practices from the digital morph as it figures in two films: *Terminator 2: Judgment Day* (1991) and *Heavenly Creatures* (1994).

These two films oppose each other in the primary phenomenological variation between figure and ground that, according to Jean Piaget, is among the first perceptual discriminations made by the developing child. Insofar as change is relative, either something changes against the world, or the world changes against something. The effective morph requires the relative, oppositional stability of either figure or ground. In *Terminator 2* it is always an object that morphs against a nonmorphing environment. *Heavenly Creatures,* however, privileges the opposite situation, in which an environment morphs around objects that do not. In *Terminator 2* the ability to change bodily form at will is the exclusive and defining property of a particular form of cyborg technology: the Cyberdyne T-1000 Terminator. The morph is not merely *what happens to* the T-1000 but is demonstrative of its essential nature as symbolized in the liquid-metal metasubstance to which this entity habitually returns between transformations. Whereas *T2* confers the plasticity of the mind to the objective body of the T-1000, *Heavenly Creatures* conversely uses the morph to objectify a shared imaginative space. Like a dream sequence, the morphs of *Heavenly Creatures* provide an extreme example of subjective cinema. Unlike *T2,* which introduces a radically different being into the mundane world, *Heavenly Creatures* uses the morph as passageway for ordinary humans to enter a heterocosm that oscillates between imaginary space and parallel universe. Although in a sense diametrically opposed, both films share a common commitment to using their morphs not to constitute gratuitous spectacles but to form *micronarratives* of their larger structural ambitions. It is this point, alongside filmic references, visually and verbally (respectively) to the fourth dimension, that distinguishes both *T2* and *Heavenly Creatures* as paradigms for analysis.

At this point, I would like to search backward for the morph's conceptual prehistory in the form of the tesseract. "I shall just sit down for a moment and pop on my boots and then I shall be on my way. Speaking of ways, pet, by the way, there *is* such a thing as a *tesseract.*"[1] It was in this sentence of Madeleine L'Engle's now famous children's book *A Wrinkle in Time* that I first came across the word "tesseract" sprinkled like fairy dust from the mouth of the mysterious Ms. Whatsit, who later becomes the subject of a profound morphological transformation:

> "Now, don't be frightened, loves," Mrs. Whatsit said. Her plump little body began to shimmer, to quiver, to shift. The wild colors of her clothes became muted, whitened. The pudding-bag shape [her body] stretched, lengthened, merged. And suddenly before the children was a creature more beautiful than any that Meg had ever imagined, and the beauty lay in far more than the outward description. Outwardly Mrs. Whatsit was surely no longer a Mrs. Whatsit. She was a marble white body with powerful flanks, something like a horse but at the same time completely unlike a horse, for from the magnificently modeled back sprang a nobly formed torso, arms, and a head resembling a man's, but a man with a perfection of dignity and virtue, an exaltation of joy such as Meg had never before seen. . . . From the shoulders slowly a pair of wings unfolded, wings made of rainbows, of light upon water, of poetry.[2]

In L'Engle's book, the tesseract is described both as a technique that enables a transmutation of form (as in the quote) and as the ability to spontaneously alter one's coordinates in space and time. This second use of the tesseract is specifically reflected in the book's title, *A Wrinkle in Time.* In this second sense, *tessering* is the act of wrinkling the fabric of reality so that great expanses of space-time may be traversed as if leaping across the top edges of a fold. So here we are given two examples of "tessering" as space-time modulation that apply neatly to both *Terminator 2* (enabling characters to alter their corporeal form) and *Heavenly Creatures* (transporting characters into different dimensions).[3]

I had never heard of the term "tesseract" before, nor would I hear of it again until a few years ago when I was reacquainted with it through the writings of experimental filmmaker and theorist Hollis Frampton. This time the tesseract emerged among things more mundane, yet it gave to the persistence of form-through-time a sense of the more extensive continuity wherein things properly reveal themselves as processes, durations. More than a century before the advent of digital imaging, the nineteenth-century emergence of photography already portended radically different ways of perceiving objects in relation to space and time. Hollis Frampton writes:

> Muybridge, in some of his earliest landscape work, seems positively to seek, of all things, waterfalls; long exposures of which produce images of a strange, ghostly substance *that is in fact the tesseract of water:* what is to be seen is not water itself, but the virtual volume it occupies during the whole time-interval of the exposure. It is certain that Muybridge was

> not the first photographer to make such pictures; my point is that he seems to have been the first to accept the "error," and then systematically to cherish it.[4]

The error that is here inextricably tied to the assertion that this was *in fact* the tesseract of water appears at first to be no more than the photographic blur: the most aleatory defect of an unsteady camera or quickly moving object. However, another passage of Frampton's elaborates on the nature and significance of this "error" and hints at the aesthetic possibilities of its strategic cultivation.

> Marcel Duchamp is speaking: "Given: 1. the waterfall; 2. the illuminating gas." (Who listens and understands?)
>
> A waterfall is not a "thing," nor is a flame of burning gas. Both are, rather, stable patterns of energy determining the boundaries of a characteristic sensible "shape" in space and time. The waterfall is present to consciousness only so long as water flows through it, and the flame, only so long as the gas continues to burn. The water may be fresh or salt, full of fish, colored with blood; the gas, acetylene or the vapour of brandy.
>
> You and I are semistable patterns of energy, maintaining in the very teeth of entropy a *characteristic shape in space and time.* I am a flame through which will eventually pass, according to Buckminster Fuller, thirty-seven tons of vegetables . . . among other things.[5]

Within the language of geometry, the tesseract is defined as a *hypercube:* the four-dimensional extension of a normal three-dimensional cube possessed simply of height, width, and breadth. However, as exemplified in the quotes from Frampton, the term "tesseract" has historically taken on a broader definition as the four-dimensional extension of any quotidian three-dimensional object, of which the hypercube stands as a primary geometric demonstration. The fourth dimension literally implies a *fourth direction* in addition to the *x, y,* and *z* axes that give our 3-D world volume beyond the 2-D plane. Thus the central ontological problem of all attempts to visually represent the tesseract must confront the central ontological and aesthetic problem of how this fourth direction might be represented *within* the three-dimensional world. All of the attempts that will be addressed here make use of a common heuristic known as *elementary parallelism* to explore the question of where this fourth direction might point.[6]

The basic principle of elementary parallelism is that consideration of

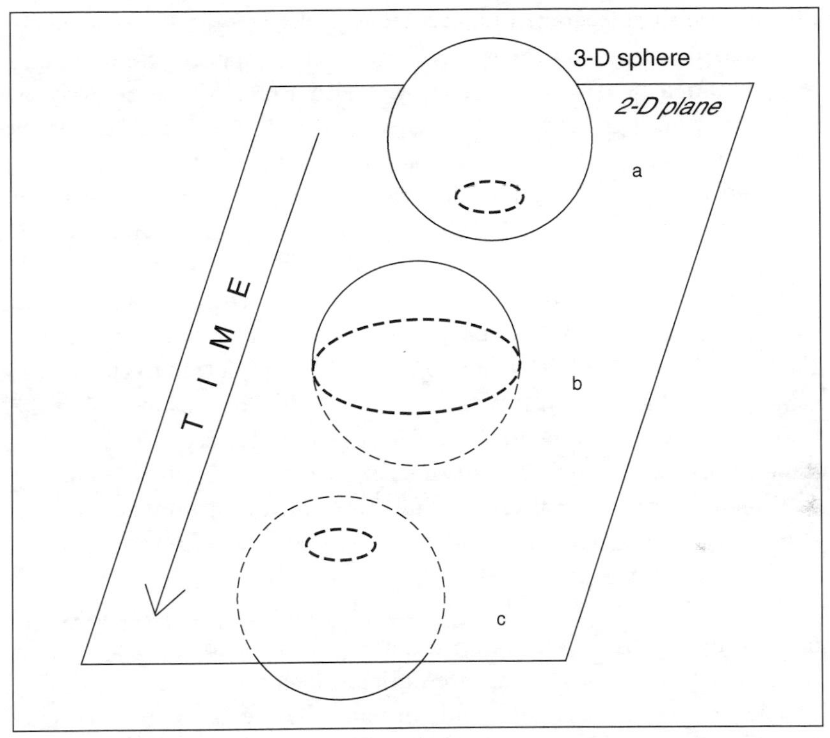

Figure 6.1 Three phases of a 3-D sphere's passage through a 2-D plane. In *a*, the sphere is just beginning to intersect the plane from above; in *b*, it is halfway through; and in *c*, it has passed almost completely through. To a 2-D entity situated within the plane, the sphere would appear only as a mysteriously growing and shrinking circle, represented by the thick broken line in each static phase. The thin broken line represents the portion of the sphere hidden from visibility beneath the plane. Courtesy of the author.

interrelationships between consecutive lower dimensions, such as, for instance, that between the third dimension (a sphere) and the second dimension (a plane), provides a key to understanding how the fourth dimension might be situated in relationship to the third. Elementary parallelism holds that any increase of dimensionality must express certain constant relational features to the lower dimension exceeded. Take, for instance, the situation of a sphere passing perpendicularly through a flat plane (see figure 6.1). When it first touches the plane, the sphere will appear as a point, and then as a series of progressively growing and shrinking circles that finally return to a point as the sphere disappears completely beneath the plane. From the point of view of the two-dimensional plane, the successive phases of the sphere's intersection appear as momentary fragments of a higher-dimensional continuum that can be inductively totalized only through the passage of time; in this case, the time taken for the sphere to pass completely through the plane from pole to pole. Extrapolated from the second and third dimension to the hypothetical encounter between the third and fourth, the 4-D *hypersphere* might appear first as a point in space (where the fourth dimension initially *touches* the third) and then as a steadily growing sphere that finally returns to a point and vanishes. In the spirit of Frampton's previous remark, "the characteristic *shape*" of this process "in space and time" may be referred to as "sphericality," but what we perceive as a growing and diminishing thing are three-dimensional moments of the progressive intersection of a four-dimensional continuum. It is in this sense that the relative persistence of three-dimensional objects through time has been understood as a necessary starting point for understanding their extension in a four-dimensional continuum.

The "shape" of this extension will be the central focus in my development of the tesseract as metaobject for the morph, and the issue of how such a four-dimensional continuum might be represented or manipulated in the three-dimensional world animates the prehistory of the digital morph and of other protomorphic time-based manipulations of objects and environments.

It may seem strange that I have used a concept that fundamentally concerns *the continuity of things through time* as the conceptual backdrop for understanding a process of radical transformation *between objects in time-based media.* In one way, the progress of my argument follows the historical development of the tesseract within twentieth-century culture. This development has two tiers. The first involves the conceptual recognition of higher-dimensional spatiality first through the notation of mathematics and then as an object of various representational strategies. The second involves the commutation of the tesseract from an *object of representation* to the *structural operating principle* behind specific technologies of representation, namely, cinema. Very quickly upon the inception of cinema, the operationalization of the tesseract within the apparatus created plastic possibilities for the medium that not only recapitulated the feats of magicians and the parlor tricks of pseudospiritualists but also suggested possibilities of transformation hitherto unseen in other media.

The conception of the tesseract as a mathematical and visual object is a product of the confluence of n-dimensional and non-Euclidean geometries. These two complementary, and often conflated, geometries became crucial components from the nineteenth century onward in a sustained attack on the Newtonian view of reality, whose demise would finally culminate with Einstein's *General Theory of Relativity* (1904) and the advent of quantum mechanics in the early twentieth century. A central component of Newtonian mechanics increasingly under attack during the nineteenth and early twentieth centuries was the understanding of space and time as totally independent of each other. Newton understood space as absolute, which meant that its existence was independent of, yet presupposed by, the existence of any objects within it. For Newton, time was something that *happened* to the physical universe, rather than a fundamental correlate of its being. Time was understood as somehow like a wind that blew on things from the outside. However, Newton's view of space, the objects within it, and time as independent of each other created special problems for the explanation of *forces.*

For Newton, there were only two types of force. Most common was the force of *displacement,* whereby one object changed the speed and vector of another through direct physical contact. A second, special category was reserved solely for the force of *gravity,* which was problematic because it acted *at a distance,* without any contact between bodies. Although Newton developed elaborate equations for measuring gravity, he offered no explanation for how or why it was actually transmitted. A recurring critique of Newton's theory of gravity concerned the fact that gravity functioned even in a vacuum, thus implying a seemingly magical transmission of force without a medium.[7] This problem was exacerbated during the nineteenth century through the growing interest in electricity and magnetism as additional "forces at a distance." Electricity and magnetism posed a problem similar to that of gravity, for they too operated in a vacuum, seemingly without any mediation. In attempts to preserve Newton's model, many scientists hypothesized that the presence of these forces proved that no vacuum is truly empty, and that the whole universe must be filled with a subtle vapor to which was given the name of *ether.* All attempts to experimentally prove the existence of an ether failed, yet at the same time, a young Austrian physicist named Georg Bernard Riemann was approaching the problem of electromagnetism from a radically different angle.

It was, in fact, a fortuitous coincidence that during the 1850s Riemann happened to be developing his own theories of higher-dimensional space while assisting his mentor Wilhelm Weber with experiments concerning the relationship between electricity and magnetism. Although others had suggested the existence of non-Euclidean geometries and sought to express higher dimensions mathematically, Riemann was the first to synthesize these areas of inquiry into a powerful and startlingly different understanding of physical reality.[8] Riemann was well aware that perennially accepted axioms of Euclidean plane geometry, such as "parallel lines never cross" and "the sum of the interior angles of a triangle always equal 180 degrees," could be easily counterdemonstrated on the condition that the plane be curved rather than flat.[9] Yet Riemann was the first to consider the implications should our 3-D world be curved in some *elementarily parallel* way to the post-Euclidean plane (see figure 6.2).

As the possibly apocryphal story goes, one day Riemann was wadding up a piece of paper covered with equations in preparation for its disposal when he suddenly asked himself: "How would it be to live in a universe that was contoured like this crumpled sheet?" Via elementary parallelism, Riemann postulated that Newton's assumption of three-dimensional space as absolute and uniform was analogous to Euclid's assumption that the geometer's plane was perfectly flat. Riemann's innovation was to suggest that the activity of "forces at a distance" could be explained if 3-D space were curved within an unseen fourth dimension, just as the 2-D plane of non-Euclidean geometry is curved within the third. To visualize the situation, Riemann described creatures living within the confines of a crumpled two-dimensional plane. Such entities would have no way of directly perceiving the curvature of their world (since this would require a three-dimensional perspective), but this curvature would be indirectly experienced in the form of mysterious, unseen "forces" that would pull objects in the direction that the plane was curved *down* into the third dimension. According to Riemann, our three-dimensional universe is likewise curved within an unseen fourth dimension whose contours are evidenced only in the activity of gravitational, electromagnetic, and, later, nuclear forces. It is from this analysis that Riemann developed his famous equation: FORCE = GEOMETRY.

With simple elegance, Riemann's equation expresses the forces that so vexed Newtonian mechanics as the epiphenomena of curvatures in three-dimensional space. Riemann further theorized that three-dimensional space is "bent" through the intersecting force of a four-dimensional continuum (just as a pin deforms the surface of paper as it punctures) whose temporally sliding intersection points constitute the objects of our universe. Riemann's model refutes Newtonian mechanics on several levels.

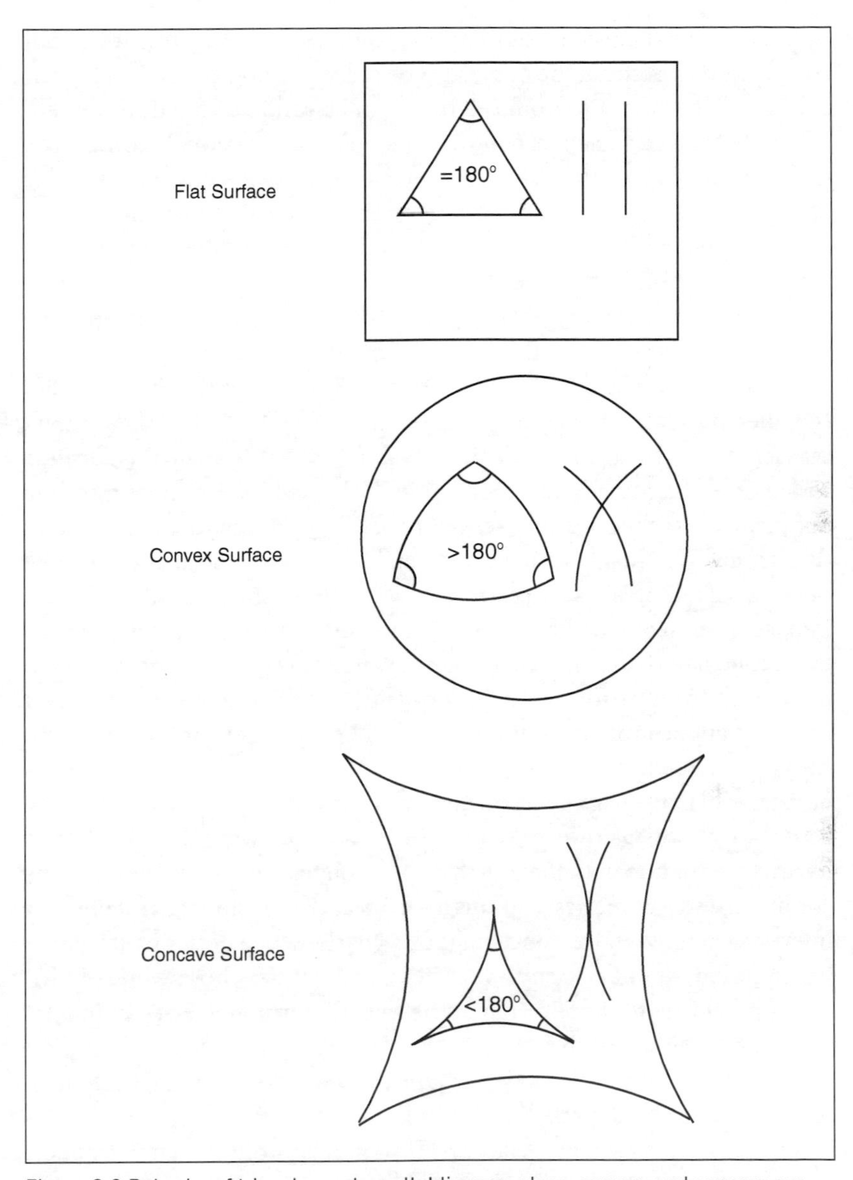

Figure 6.2 Behavior of triangles and parallel lines on plane, convex, and concave surfaces. On plane surface *a,* all the axioms of Euclidean plane geometry apply: the sum of the angles of an equilateral triangle always equal 180 degrees, and parallel lines never cross. On convex surface *b* and concave surface *c,* however, the sum of the sides of an equilateral triangle will equal more or less than 180 degrees given any degree of curvature to the surface, and parallel lines will always cross. Courtesy of the author.

First, space is not absolute but a fabric contoured by the presence of objects within it. Second, the curvature of space (analogized through non-Euclidean geometry) is produced by the intersection of the third dimension with four-dimensional continuums (via n-dimensional geometries). Third, it is through time that the geometry of space changes, and the shape of space cannot be separated from the objects within it and the passage of time. In his famous oral presentation of 1854, Riemann demonstrated how the fourth dimension could be represented mathematically by a simple extension of the Pythagorean theorem. For three-dimensional space, the theorem states that the sum of the squares of three adjacent sides of a cube is equal to the square of the diagonal; so if *a, b, c* represent the sides of a cube and *d* its diagonal length, then $a^2 + b^2 + c^2 = d^2$. Again, via elementary parallelism, Riemann was able to extrapolate this formula to the case of a four-dimensional cube, or hypercube, simply by adding another term to the equation (see figure 6.3).[10] Although Riemann was not directly involved in producing visual representations of four-dimensional objects, his mathematical description of the hypercube would have great impact on subsequent attempts to describe and render it visually. Through the second half of the nineteenth century, attempts to represent the fourth dimension were performed largely outside the parameters of any direct scientific application. In fact, the most avid proponents of the fourth dimension were clergymen, mystics, and artists who saw in it a scientific basis of explanation for ghosts, clairvoyance, the afterlife, and paranormal activity of all sorts. However, within the decade following Riemann's 1854 lecture, fascination with the fourth dimension had spread throughout the continent and to America. Although the breadth of influence and popular interest generated by the concept far exceeds the scope of this paper, here I will spotlight specific scientists, writers, and artists whose work significantly prefigures the application of the four-dimensional tesseract in relation to the digital morph.

Charles Howard Hinton is perhaps the best known of those attempting to visualize the fourth dimension in the late 1800s. A mathematician by training and mystic by vocation, Hinton first published on the subject in England in 1880 and by the end of the decade had moved to the United States, where he taught mathematics at Princeton University (1893–1897) and the University of Minnesota (1897–1900). During this period, Hinton published several articles on the fourth dimension, which were anthologized first in 1884, and again in 1895, in a volume entitled *Scientific Romances*.[11] In 1884 Hinton elaborated the evolutionary and mystical implications of the fourth dimension in *A New Era of Thought*, and in 1904

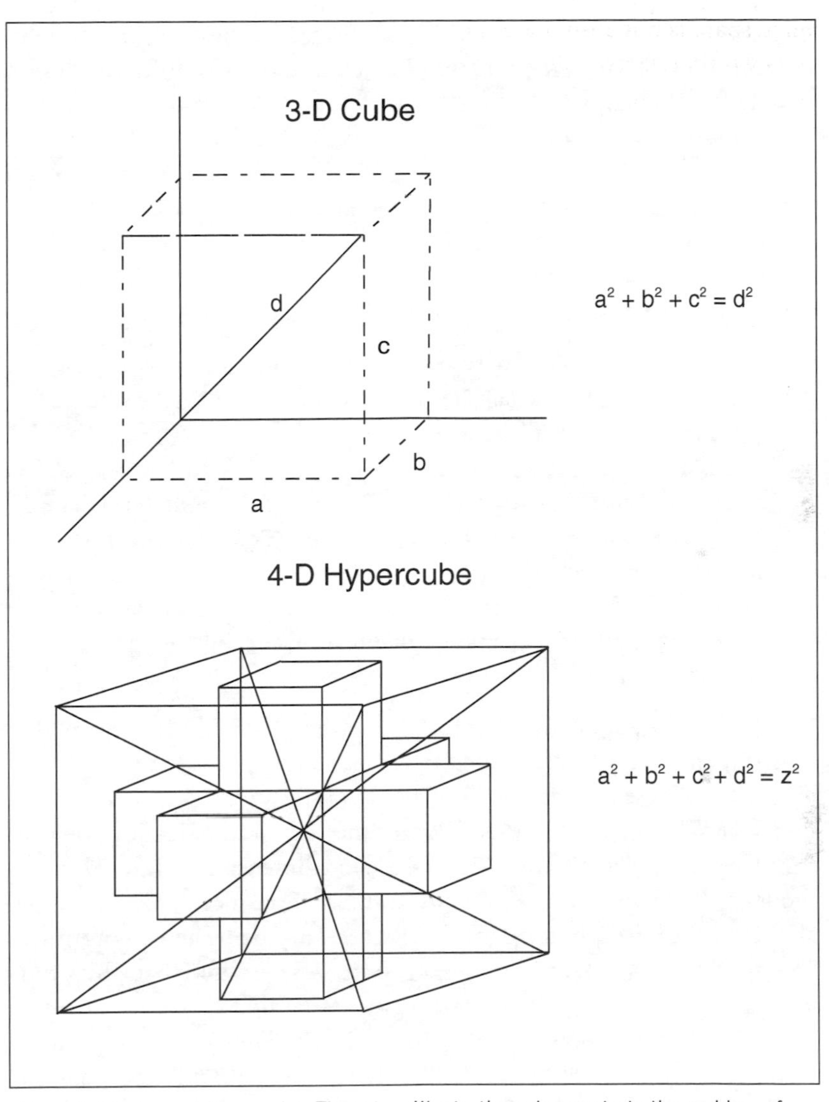

Figure 6.3 Cube and hypercube. These two illustrations demonstrate the problem of visualizing 4-D objects despite the ease of their mathematical notation. The first shows the Pythagorean equation $a^2 + b^2 + c^2 = d^2$ applied visually to a 3-D cube. The second shows a simulated hypercube that attempts to express the equation visually: $a^2 + b^2 + c^2 + d^2 = z^2$. The attempt to visualize the volume represented by the equation produces an optical illusion: because diagonal *z* literally points in a direction perpendicular to all the other directions in the 3-D world, the hypercube seems to rotate as the figure leads the eye to search for 4-D diagonals everywhere but ultimately allows the eye to fix nowhere. Courtesy of the author.

he published *The Fourth Dimension* as a grand synthesis of his thought. Hinton's work is significant for understanding the scientific revolution initiated by Riemann as the harbinger of a *perceptual revolution* through which human perception would *open up* onto the frontier of higher dimensionality. To this end, Hinton devised a number of techniques to aid in the development of our perceptual faculties through the exercise of the neglected organ of the mental, or inner, eye. Hinton sums up his project in *The Fourth Dimension* as follows:

> Here, for the first time, the fact of the power of conception of four-dimensional space is demonstrated, and the means of educating it are given. And I propose a complete system of work, of which the volume on four space is the first installment.
>
> I shall bring forward a complete system of four-dimensional thought—mechanics, science, and art. The necessary condition is that the mind acquire the power of using four-dimensional space as it now does three-dimensional.
>
> And there is a condition which is no less important. We can never see, for instance, four-dimensional pictures with our bodily eyes, but we can with our mental and inner eye.[12]

Via elementary parallelism, Hinton conceived of a number of ways that four-dimensional objects might be visually conceptualized in the third dimension. These involved extrapolations of the means by which a three-dimensional cube could be unfolded onto, cast its shadow on, be rendered in perspective, and intersect the second dimension. Within Hinton's thought, the tesseract came to stand for the transcendent object to which all these techniques pointed without, by necessity, being able to capture it. The most important technique for our purposes here, and that which borrows most directly from Riemann, is *cross-sectioning*, the simplest example of which is the preceding situation of the sphere passing through the plane (see figure 6.1). Hinton developed a series of colored cubes that, like the growing and shrinking circles produced by the intersection of the 3-D sphere and 2-D plane, stood, in the manner of 3-D segments, for the abstracted instances of a 4-D hypercube's passage through the three-dimensional world.

Both Riemann's and Hinton's analogies produce time-oriented visions of a four-dimensional body. Thus it is in the physics of Riemann and the visualizations of Hinton that the previously separated categories of space and time conjoin into a unified category of *space-time*. Here it is important to emphasize that time is not itself the *equivalent* of the fourth dimension, but rather that the tesseract reveals itself within the third dimension only through time's passage just as the 3-D sphere only reveals itself as larger and smaller circles through its temporally protracted intersection with the 2-D plane.

Beyond the United States, Hinton's writings also had direct and indirect effects on aesthetic movements in Europe at the turn of the century. In France, Hinton's work was read by the mathematician E. Jouffret, whose *Traité élémentaire de géométrie à quatre dimensions* is reputed to have had a significant impact on the Puteaux Cubists group, which included Marcel Duchamp. As an artist avidly interested in the fourth dimension, Duchamp was directly familiar with Hinton's ideas and also knew of them through a number of popular glosses including the essays of Gaston de Pawlowski and the writings of the American Claude Bragdon, who, after Hinton's death in 1907, became the best-known writer on the fourth dimension in the United States. As I have already suggested, Hinton was one among many thinkers who expressed several different means of visualizing the fourth dimension, not all of which took temporality as a primary component. For instance, of prime interest to the cubist group was how a four-dimensional perspective might be visually expressed as a simultaneity of differently situated three-dimensional perspectives. This idea was the driving force behind the characteristically cubist trope of *faceting*, by which the perspectivally exclusive sides of a three-dimensional object or scene are unfolded onto one visual plane.[13] Duchamp's *Nude Descending a Staircase* (1912) broke with prevailing cubist practice by compressing time onto the canvas rather than unfolding space. The temporally distended figure within Duchamp's image is actually a clear recollection of futurist surfaces, as the machinic contours of the "nude's" body seem to indicate the underlying geometric nature of the four-dimensionally extended figure. Duchamp's "nude" thus represents an early attempt at rendering the tesseract as an element of space graphically protracted by its passage through time.

Although Duchamp's representation of the tesseract as a space-time continuum draws theoretically from the mathematics of Riemann, the visual strategy employed pays homage not only to Hinton but also to the protocinematic work of Eadweard Muybridge and Jules Etienne Marey. Like the waterfalls that Frampton identifies in Muybridge's early landscape photography, the *nakedness* of Duchamp's figure corresponds to a similar blurring effect that distorts the features of three-dimensional thingliness in favor of the representation of a virtual volume expressive of a characteristic shape in space-time. Duchamp's is a new type of "nude," which undergoes a more primordial disrobing from the perspective of the fourth dimension. Duchamp's nude also bears the analytic intentions of Muybridge's later *Studies in Human and Animal Locomotion,* which sought less to capture the tesseract as a graphic object than to analyze movement within space-time as instances abstracted out of process.

Marey was not expressly interested in the fourth dimension. Rather, he sought to analyze movement scientifically. In what he called his "chronophotographs" (begun in the 1880s), Marey would mark his human subjects with white dots set at particular plot points on their bodies (knees, elbows, fingers, etc). The subjects were then instructed to complete some simple movement such as a forward jump, for whose duration the camera shutter would be left open. This graphic-synchronic rendering of diachronic movement reveals a segment of the tesseract whose characteristic shape is isolated by the positioning of the dots as formal constraints designating the limits of the human body and the contours of its progression.

Muybridge and Marey are significant insofar as they exploited the ability of photographic-based technology to express time and space each as functions of the other. In straddling the divide between the protocinematic and the cinematic, they also stand as pivotal figures in the transition of the tesseract from an object of representation to the operating principle of cinema as emerging representational technology. The invention of cinema enabled the mechanical disassembly of the tesseract into a series of static frames and its subsequent reanimation through the redundant and invisible activity of the apparatus. Once the space-time of the day-to-day world could be reanimated through the camera and projector, the challenge quickly became how to manipulate the nature of objects in the interest not only of the dramatic but also of the fantastic. It is through attempts to maximize the plastic possibilities of cinema that the tesseract returned to the cinema not only as mechanized operating principle but as a sculptural tool attempting to assert complete control of the object in duration. I think it is in this sense fitting that I turn to the writing of a sculptor to bring the figure of the tesseract and the technology of morphing into contact.

The sculptor Oscar Dominguez, who was involved with both the cubists and the surrealists, makes reference in his 1942 essay "La Pétrification du temps" to a new type of surface emerging within modern art to which he gives the name *lithochronic:*

> Let us imagine for a minute any three-dimensional body, an African lion for example, between any two moments of its existence. Between the lion L(0), or the lion at the moment t = 0, and the lion L(1) or final lion, is located an infinity of African lions, of diverse aspects and forms. Now if we consider the ensemble formed by all the points of lion to all its instants and in all its positions, and then we trace the enveloping surface, we will obtain an *enveloping super-lion* endowed with extremely delicate and nuanced morphological characteristics. It is to such surfaces that we give the name *lithochronic*.[14]

Imagine Dominguez's lithochronic surface extending from lion L(0) to lion L(1). Now imagine that we retain L(0) but substitute lion L(1) with the figure of a naked man, so that the "nuanced . . . morphological characteristics" now span the *graphic* distance between the two different objects instead of, or in addition to, one object's movement through space-time. This is essentially the principle behind the digital morph: between any two or more objects (still or moving) may be generated a *lithochronic surface* that graphically connects the two objects and, in that connection, travels the formal difference between them through time.

All morphing programs, including the most common, Elastic Reality, use a visualized *tesseract* as a tool for controlling the transformations of a lithochronic surface from a "source" to a "target" image. The "source" and "target" image are displayed side by side and are connected to each other by a series of "splines." In one particular interface window of Elastic Reality, for example, these splines are seen connecting closed two- or three-dimensional shapes: these lines may connect a mouth to a mouth, a head to a head, or any delimitable area on the body of one figure to any area on another. The splines define the trajectories across which the values of the "source" image literally slide toward those of the "target" image. These computergraphic programs make it apparent that the problem of the morph is not primarily an issue of discontinuity. Indeed, quite the opposite: we are struck by the direct representation of two apparently discontinuous beings or things sharing in a continuous material substratum. The measure of this continuity is witnessed in the unbroken-ness of the tesseract. The tesseract is always a continuum, but it need not necessarily be occupied by the *same* object at both ends.

It is this unique deployment of the tesseract, as continuous substratum for a plurality of objects, that distinguishes the digital morph from attempts at transformation within purely analog cinema. The central difference is in fact a *difference at the center* (or apex) of transformation between the "source" and "target" of the digital morph. Within any morph between two objects there is a midpoint at which the morph is minimally recognizable as either "source" or "target" image. It is at the moment of midpoint that, if only just for an instant, the morph lapses from the order of known things. Most important, this lapse (or lack) of formal definition is still figured in full three-dimensional extrusion, and the paradoxical presence of *being-without-thing-ness* blinks at us like some denuded metasubstance stripped of the overdetermined trappings of symbolic designation and fixity.

At this crucial apex of underdetermined transformation, the digital morph remains visibly extruded in three-dimensional space. However, the dissolve, so often used to effect transformation within analog cinema, invariably collapses back into two dimensionality, and this transformation is obscured. An early example of the dissolve—used as protomorph—may be found in George Waggner's *The Wolf Man* (1941). The film uses a slow dissolve between static frames of the actor in various stages of make-up to effect the transformation of man into monster and vice versa. However, the requirements of this superimposition invariably compromise elements that distinguish the cinema from still photography. Most importantly, the image must retain a static tableau-framing through the duration of the transformation so that all other differences between the successive frames can be minimized or, ideally, eliminated. Unlike still photography, moving images actualize the intentional space of the image, implying a thickness to things, transforming the human face from Barthes's death mask[15] into Balázs's living *microphysiognomy.*[16] That is, intentionality is not simply a static form but only becomes apparent along a directional vector *as gesture* (however small). The still photograph freezes these vectors, thereby muting the intentional activity that characterizes cinema. Through the transformation of man into monster, the figure loses its dramatic thickness and collapses toward photographic still life. A corresponding limitation of the dissolve is the inability to control every element of change from frame to frame. Remember that between Lion(0) and Lion(1) Dominguez's lithochronic surface is made up of an infinite number of lions that represent all the positions of a movement through space-time. As cinema has demonstrated, all of these intermediary lions may be removed, save for eighteen to twenty-four per second, and the graphic differences between each position will not be detected by the naked eye. Although the cinematic dissolve, as featured in *The Wolf Man,* aspires to the plasticity of animation in which every detail of every frame is under the total control of the filmmaker, the commitment to photo-realism invariably produces observable discontinuities between the "source" and "target" images. The superimposition of the photographic planes fractures the illusion of depth, drawing attention to the two-dimensional surface of the image. At the crucial moment of organic transformation, the thing (the wolf man, in this case) suddenly appears robbed of volume, inorganic, flat. The effects of photomontage pronounce themselves by default in the subtle, but recognizable, skewing of perspectival cues, lighting values, and double exposures, which all constitute fractures in the lithochronic surface of the tesseract. At the apex of transformation, the dissolve remains indexical but compromises the illusion of three-dimensionally extruded life that characterizes cinema. Unlike the digital morph, which produces *being-without-thingness,* the cinematic dissolve produces mere *thingness-without-being.*

To better understand the differentiation of the digital morph from the cinematic dissolve, it will be useful to examine some concrete examples of morphing in *Terminator 2* and *Heavenly Creatures.* Through an analysis of these two films, I will also explore the crucial role that digital technology plays in producing the morph as a novel cultural artifact. As I have already suggested, the novelty of the digital morph resides in its smooth manipulation of the tesseract beyond the limitations of the analog dissolve. The crucial difference here lies at the structural-material level of the two media, and is reflexively pointed to within these films' use of the morph.

Terminator 2 pits the original Terminator cyborg (*The Terminator,* [1984]) against the new "liquid metal" T-1000. Following up on the narrative of the first film, we learn that the original Terminator had been captured by the human resistance forces of the future, reprogrammed to protect human beings, and then returned to the past. At stake is the life of John, the adolescent son of Sarah Connor, who in the future will become the leader of human resistance against tyrannical, misanthropic computers. (The copresence of characters from the present and future is itself enabled by something akin to the *tessering* that Mrs. Whatsit describes as *A Wrinkle in Time:* however, this space-time modulation is not directly represented through a morph.)

The two rival Terminators not only represent the forces of good and evil but also structurally allegorize the tension between the analog-machinic and the digital within technologies of representation. Specifically, the T-1000 embodies the transcendence of the digital over the limitations of the analog cinematic apparatus. Both terminators are copies—*simulacra*—but each ultimately emerges as a very different species of representation. The original terminator is a *copy* of a human being, but the T-1000 is a *copier:* a shape-shifter. By similar comparison, the photographic cinema photochemically and mechanically represents objects of which it becomes a fixed copy, whereas the digital represents pixels, which can be made to copy anything. Unlike the composite of analog photographic cells that are individually "circles of confusion," beyond the control of the photographer, the digital provides total control of the values ascribed to each and every pixel. In this way, the digital image is in no way fixed to the things it represents, just as the T-1000 transcends all the manifestations that it temporarily assumes.

Whereas the photographic has only a *molar* relationship to its object, the digital has total control at the *molecular* level of the image. Instead of the fundamental confusion of a photochemical process, the digital can reduce all *qualia* of the image to the clarity of a computation. Indeed, I would suggest that the possibility of higher-dimensional convolutions of space within digital media is a direct result of its atomic reduction of *moving image* to *moving algorithm.* The digital commutes Riemann's equation FORCE = GEOMETRY to the substructural level of the moving image.

Through its embodiment of the digital, the tesseract that the T-1000 carves through space-time becomes totally plastic to its will. The T-1000 has the ability to alter not only its positional form but also the structural components of its body, and it can change its body as capriciously as we can change our minds. The T-1000's transformations perform violations of human physiology that relegate its physicality to the epiphenomenon of some causal source emanating from a dimension supplementary to those of the visible world. Not only does the T-1000 turn into its victims, but it also has the ability to turn inside out within its own body, so that its face appears through the back of its head. It is worth noting that this particular maneuver has a rich lineage in terms of attempts to represent the *rotation* of a three-dimensional object into the fourth dimension. Elementary parallelism indicates that an element of n dimensions always rotates around an axis of $n - 1$ dimensions within a space of $n + 1$ dimensions. For instance, a line (of one dimension) always rotates around a point (of no dimension) on a plane (of two dimensions). A plane (of two dimensions) rotates around a line (of one dimension) in a space (of three dimensions). The next step would suggest that in an unknown hyperspace (of four dimensions), a volume (of three dimensions) would rotate around a plane (of two dimensions). It is simple enough to conceive of how a three-dimensional object rotates around a line within three-dimensional space, just as a plane rotated around a point remains only in two-dimensional space; but the mind reels at how an object might *rotate* around a plane into the fourth dimension.

Early theorists of the fourth dimension such as Hinton (and later, in the early twentieth century, Claude Bragdon) saw the mirror as a virtual means of performing a rotation into the fourth dimension otherwise not possible within the third:

> In the mirror image of a solid we have a representation of what would result from a four-dimensional revolution, the surface of the mirror being the plane about which movement takes place. If such a change in position were effected in the constituent parts of the body as a mirror image of it *represents,* the body would have undergone a revolution in the fourth-dimension.[17]

For instance, there is no maneuver that can be performed in three-dimensional space that will switch the sides of one's face so that the right eye is in the place of the left and vice versa, except in the reflected image of the mirror. Duchamp referred to the mirror as a "hinge plane,"[18] and Hinton found in the mirrorlike bilateral symmetry of living bodies proof of a fourth dimension spilling out into the world through a central seam that defines all biological life. Another type of rotation that mimics the reversal of the mirror is that of the glove turning inside out, which succeeds in changing the right hand into the left. This is in fact most closely akin to the T-1000's performance as it literally reverses its front and back across a virtual plane situated somewhere in the middle of its body. Thus able literally to mirror itself in rotation around a plane, the T-1000 is also mirrorlike in terms of its propensity to turn into its victims, and in terms of the reflective liquid-metal flux to which it returns between transitions.

The T-1000's ability to turn into many different figures and, in one scene, to be shattered into a million pieces without compromising its fundamental continuity, is instructive of another property of elementary parallelism: that every object of n dimensions is bounded by objects of n - 1 dimensions. Thus a line is bounded by points, a plane is bounded by lines, a solid is bounded by planes (flat or curved), and a four-dimensional object must be bounded by three-dimensional solids. Bounded by a variety of three-dimensional bodies but not restricted to any one (just as a cube is bounded by planes but not contained within any one of its sides), the T-1000 can be frozen and broken into a million pieces, only to reassemble miraculously via some type of distributed memory. This automatic re-assembly is perhaps most instructive of the fact that the T-1000 is unified in a dimension beyond its visible body. One recurring "flatland" analogy is helpful for understanding this phenomenon. If the fingers of a three-dimensional hand fell on the plane of flatland, those living on the plane would perceive five different objects and be unable to conceive how, in the third dimension, these objects were connected to the agency of a single hand.[19] Similarly, what appear within the third dimension as discontinuous things, exhibiting a strange organizing principle among themselves, might be the "fingers," so to speak, of a four-dimensional hand.

The nature of the T-1000 to be bounded by, though not restricted to, different three-dimensional bodies as so many "facets" or sides of one unified source, is dramatically expressed in the T-1000's final demise. It seems that the Achilles' heel of the T-1000 resides not in discontinuity (over which it displays clear mastery) but in the dissipating power of a more fundamental continuity. Much like the Wicked Witch of the West, who melts when splashed with water, the T-1000 is returned to a more primary form of nondifferentiation when it is *alloyed* in a vat of molten metal. In the midst of its death throes, the T-1000 leaps repeatedly from the molten metal to perform a repertoire of the various extrusions that it has assumed during the course of the film. Its being as an object within our "order of things" is thus figured ultimately as performative act. Before finally diffusing into pure molten flux, the T-1000 appears as an orb whose surface peels away as it spills itself forward from its center. The effect of this shot suggests a Möbius strip, with inside and outside connected in one continuous loop. In n-dimensional geometries, this movement describes a *pseudosphere.* The pseudosphere is a theoretically 4-D object that is represented by a sphere whose two halves are imploded toward its center of gravity and across the plane of the sphere's circumference, their leading edges describing asymptotic arcs toward infinity. Any digital artist with a background in mathematics would understand this object, and I would argue that its use within this film has reflexive and iconic significance. It is worth remarking that several of the higher-dimensional figurations explored here present the fourth dimension as being a transcendent space, in addition to a lithochronic protraction of ordinary space-time. Here, I am not arguing for the preeminence of any particular articulation of the nature of higher dimensions, although I do want to suggest that it is the digital mastery of the tesseract as space-time continuum that fundamentally enables these other representations of higher-dimensional processes.

But what is this metasubstance that blinks at us from the apex of the morph, and that in *Terminator 2* is hyperbolized in the quicksilver substratum of the T-1000? In his essay "Why Does the *Phallus* Appear?" Slavoj Žižek investigates the amorphous appearance of the face of monsters and deformed creatures from *The Phantom of the Opera* (1925) to *The Elephant Man* (1980). Žižek describes the characteristics of the monstrous face as follows:

> The flesh has not yet assumed definite features, it dwells in a kind of preontological state, as if "melted," as if having undergone an anamorphotic deformation; the horror lies not in his death mask, but rather in what is concealed beneath it, in the palpitating skinned flesh—everyone who catches sight of this amorphous life substance has entered the forbidden domain. . . . Therein lies the ultimate paradox of the "living dead": as if death, the death stench it spreads, is a mask sheltering a life far more alive than our ordinary daily life . . . having access to the life substance prior to its symbolic mortification.[20]

I would suggest that the psychoanalytic significance of the morph is located at the apex of its transformation in its *revelation* of this "life substance." For Žižek, this "life substance" is nothing other than the Lacanian Real; a sort of presymbolic force of extrusion, often figured as the paradoxical fecundity of dead matter. The antagonistic force of this substance, which haunts things as the excess perpetually escaping symbolization, arises from its tendency to disrupt *and* confute the *discontinuities* that stabilize the "order of things," both social and familial. In *Terminator 2,* against the protection of the human mother and the mechanical father (the original Terminator), both of whom fall on the side of the symbolic, the T-1000 appears as a force of primary extrusion, the chaotic life force that threatens to annihilate the impromptu family. However, the substratum that the T-1000 represents is not the excess fecundity of the traditional Grim Reaper, but rather the silicon-based future of a world taken over by digital computers. That this future is jointly the production of global capitalism run amok is also telling, as Žižek remarks that in one sense "capital is the Thing 'par excellence': a chimeric apparition which, although it can nowhere be spotted as a positive, clearly delimited entity, nonetheless functions as the ultimate Thing regulating our lives."[21] In a profound way, the T-1000 embodies the exchange value of capitalism and commutes its logic to a new *silicon standard* of extruded being that seeks to exterminate the old standard of carbon-based life and its labor.

The organ-ized, symbolized body that constitutes our ground for human intersubjectivity is merely an epiphenomenon of the T-1000. Intentional agency cannot be located in any of its visible organs, for they are all transitory configurations of some deeper intentional source that transcends any of the manifestations it assumes. In this sense, whatever invariants constitute the identity of the T-1000 are deferred to a dimension supplementary to that of the 3-D world. It is important here to remember that the realization of this creature *in representation* cannot be separated from the enabling technology of digital media and its ability visually to express algorithmic geometries whose dimensionality exceeds that of the quotidian world. The invisible organ-ization of the T-1000 is in a very real way commuted to the computational engine *behind* the digital image. In this way, the changing materiality of the medium must be understood as a correlate of the phenomenological and psychoanalytic significance of the novel objects it produces. Thus, as I have already suggested, the liquid-metal substratum of the T-1000 can be understood as a reflection of the differential materiality of digital media vis-à-vis analog, photographic-based cinema, and as a more concrete figuration of the deeper continuities that the morph reveals.

Although quite different in its use of morphing, *Heavenly Creatures* (1994) presents a similar metasubstance, but this time characteristic of an other dimension opened up through the agency of a profound psychic transformation. In this case, the substance is *plasticine:* a gray modeling clay out of which Juliet and Pauline fashion the characters of their imaginary kingdom Borovnia. This fantasy realm in which the two teenage girls spend all their spare time is also referred to as "the fourth world." This "fourth world" comes to consume the girl's lives as they attempt to objectify their imaginary space as a shared external reality. The girls' need to incarnate their "fourth world" is fueled by their parents' mounting disapproval of the relationship. Having no forum of expression within the visible world, Juliet's and Pauline's longing for each other implodes into the heterocosm of Borovnia, where reality is completely plastic to the force of desire. In this sense, the film's central morph seems to perform a strange twist on Riemann's equation FORCE = GEOMETRY. It is as if the force of the girls' desire to be together is substituted for the natural forces that, in Riemann's formula, bend and deform space. The morph thus appears as the agency of this desire, literally transforming space according to its requirements.

The morph in question occurs at a particularly traumatic point in the film, when Juliet's parents have decided to travel abroad, leaving her behind. Both girls are taking one of their usual walks in the fields when Juliet suddenly pauses from her tears and points off into the distance: "Look Pauline, can't you see it?" A ripple becomes visible across the surface of the landscape, and then, suddenly, beautifully manicured trees and marble fountains swell from the ground, giant butterflies spring out of thin air, and the sky turns an impossibly bright shade of blue. They both exclaim, "The fourth world!" and run off into the transformed landscape.

This morph is of particular interest because it figures the emergence of the "fourth world" within a deep-focus landscape. Yet by emerging not as or with isolated objects on the screen but through every element of environment (save for the bodies of the two girls), the depth of the landscape appears collapsed on one uniform plane in the midst of the transformation. It is not that the three dimensionality of the image is compromised, as in the earlier case of the photographic dissolve, but that the extrusions of the "fourth world" appear as perturbations within one uniform sheet on which the perspectival cues of the mundane "source" world contort and fade away. It is as if the mundane world becomes reduced to a perspective painting whose canvas is stretched and contoured according to a more vital force surging from behind. The film gives this effect the sense of total enclosure by showing the morph from three perspectives in three consecutive shots. The first shot pictures the girls from behind running head-on into the morphing environment. In the second shot, the girls run toward the camera as the space behind them swells and twists like a deforming balloon. In the third shot, the camera tracks from the side as the girls walk past upwelling landscape formations. This strange coincidence of effects, whereby the environment is both reduced to a plane and, at the same time, presented as totally enveloping, corresponds to a particularly puzzling set of features that have been associated with a hypothetical four-dimensional intersection of the three-dimensional world. The French mathematician E. Jouffret coined the term *tranche infiniment mince*, "infinitely thin slice," to account for a particular characteristic produced via elementary parallelism:[22] Jouffret observed that any line can be divided by an infinite number of points, any plane by an infinite number of lines, and any solid by an infinite number of planes. The general rule that may be deduced is that every object of $n - 1$ dimensions must appear "infinitely thin" within a space of n – dimensions. From this point, one can raise the question "what does it mean for the third dimension to be infinitely thin in relation to the fourth?" In terms of the temporal expressions of the tesseract, the "infinite thin-ness" of the third dimension to the fourth might be witnessed in the *superporosity* of everything to time. Nothing exists outside of time, yet time seems to come from everywhere and nowhere at once—just as the passage of our hands through the second dimension would appear to be acting within it, but coming from a direction invisible and perpendicular to all known directions on the plane. It is in this sense that the "fourth world" emerges seemingly from everywhere at once but simultaneously reduces the thickness and depth of the mundane world to an infinitely thin threshold through which it may pass.

The porosity of the mundane to the "fourth world" remains an issue within *Heavenly Creatures* even after the execution of this particular morph. The "fourth world" exposed by the morph evaporates when Juliet suddenly coughs up blood and, struck by a relapse of her childhood bout with tuberculosis, is forced into the isolation of a clinic. Deprived of direct contact with each other, the girls' relationship—and the film's narrative—is continued through their written correspondence, in which Pauline and Juliet adopt the personae of their characters Charles and Devora, whose despotic son Diello begins to puncture the film's shrinking domain of reality. A life-size, armored knight made entirely of gray plasticine, Diello spontaneously appears to skewer a priest at Juliet's bedside, and later to literally split in two the psychoanalyst who councils Pauline about the "unhealthiness" of her relationship with Juliet. The morph has set a 4-D seed within the film's body whose germination starts extruding the girls' imaginary realm of Borovnia into the film's mundane 3-D world. Like the T-1000, Diello enters the world of the film as the three-dimensional manifestation of a higher-dimensional source. However, the two films differ fundamentally in where they locate the agency of the morph. *Terminator 2* locates the morph as the power of a new technology, whereas *Heavenly Creatures* uses it to externalize the power of human desire on the canvas of mise-en-scène.

I have tried to demonstrate how the transformation of the tesseract from an object of representation to a representational tool has permitted the realization within visual culture of objects previously confined to the realm of mathematical and theoretical abstraction. Although the first impulse of this added capacity for representation has been to manipulate the tesseract in dramatic fashion and to produce novel phenomenal objects that reflect the unique structure of the enabling technology, the future tendency seems to lean toward an increasingly invisible use of the morph. The movement of the morph from foregrounded object in *Terminator 2* to background environment in *Heavenly Creatures* traces the beginnings of a trajectory toward a more cosmetic use of the morph whose subtle and nuanced enhancements of reality will leave these films as relics of a second "cinema of attractions." Or else, rather than being reduced to mere effect, perhaps the form, content, and contextualization of morphing technologies will find creative ways to intertwine with the narratives of their "host" films, and a new form of meaning will emerge from within the "effect" as a function of the higher-dimensional concepts that they are uniquely suited to express. A key film in this respect is *Dark City* (1998), in which the entire world of the film is morphed into existence every night by a mysterious group of "Strangers" who control the outer reality of the mise-en-scène and implant memories into the inner reality of their human subjects. The film is striking insofar as it strives to move beyond the standard metamorphic opposition between figure and ground as an entire city swells into being. The resulting conflations of figure and ground greatly contribute to the pervasive sense of alienation in the film, as well as perhaps the alienation of the audience from the film, since it is only the human figures that remain *amorphically* foregrounded against an ever-changing world. Even though the human figure remains amorphically foregrounded, traditional identification becomes impossible because the psychic continuity of the characters and the objective continuity of the film's external world as minimal context for value are perpetually and thoroughly undermined. I introduce this film only as a coda to suggest that if the morph is to retain its value and dynamism, it must engage the existential problematics that its cinematic usage entails.

Notes

1. Madeleine L'Engle, *A Wrinkle in Time* (Prentice Hall: New York, 1964), 26.
2. Ibid., 63.
3. I should point out that this division is not absolute but rather reflects a variant privileging. *Heavenly Creatures* provides its own interesting incidents of objects that morph, although they are always given as the product of mind—not technology. Correspondingly, both *T2* and the first *Terminator* begin with an act of time travel, although the future world is realistic and not fantastic. It would be interesting to see how the films' different generic commitments to fantasy and

science fiction influence their differential uses of the morph and special effects technology in general.

4. Hollis Frampton, "Eadweard Muybridge: Fragments of a Tesseract," in *Circles of Confusion* (New York: Anthology Film Archives, 1982), 76; italics mine.

5. Hollis Frampton, "A Pentagram for Conjuring the Narrative," in *Circles of Confusion*, 62.

6. The French mathematician and writer E. Jouffret first coined the term "elementary parallelism" to name a class of explanatory devices the practice of which is as old as that of geometry itself, although it was through the thought experiments of Riemann that this implication of the principle was carried out in terms of higher dimensions. E[sprit Pascal]. Jouffret, *Traité élémentaire de géométrie à quatre dimensions* (Paris: Gauthier-Villars, 1904).

7. For instance, other forces that appear to act at a distance, such as wind or the concussion wave of an explosion, are actually being transmitted through the medium of oxygen. However, the existence of gravity, even through a vacuum, troubles Newton's view that space is unaffected by the matter within it, and vice versa.

8. For more information on forerunners of Riemann in the fields of non-Euclidean and n-dimensional geometries, see Linda Dalrymple Henderson, *The Fourth Dimension and Non-Euclidean Geometries in Modern Art* (Princeton: Princeton University Press, 1983).

9. Given a plane with any degree of concave or convex curvature, the sum of the angles of a triangle inscribed on it will be more or less than 180° respectively, and parallel lines will *always* cross.

10. The extended equation appears thus: $a^2 + b^2 + c^2 + d^2 = z^2$. In this case, z stands for the diagonal of the hypercube.

11. See Charles Howard Hinton, *Scientific Romances*, 1st ser. (London: Swan Sonnenschein, 1884–1885, 1886; reprint, New York: Arno Press, 1976); *A New Era of Thought* (London: Swan Sonnenschein, 1888); and *The Fourth Dimension* (London: Swan Sonnenschein, 1904; New York: John Lane, 1904). For more on the life and work of Charles Howard Hinton, see Henderson, *The Fourth Dimension and Non-Euclidean Geometry in Modern Art.*

12. Hinton, *The Fourth Dimension*, 207.

13. Faceting aims to undermine a fundamental invariant of three-dimensionally situated embodiment, which dictates that any object of perception will be partially transcendent to the perceiver. Faceting, as in cubist painting, attempts to produce the privilege of a four-dimensional perspective upon the third dimension in an elementarily parallel way to the advantage of a three-dimensional perspective upon a plane in which the four sides of a square appear simultaneously, while the

two-dimensional inhabitant of the plane must move around the square to deduce its total shape.

14. "Litho-chronic" would seem literally to mean "time-relief." Oscar Dominguez, "La Pétrification du temps," in *Surrealists on Art,* trans. and ed. Lucy Lippard (Englewood Cliffs, N.J.: Prentice Hall, 1970), 109 (quoted in Henderson, 348).

15. See Roland Barthes, "The Face of Garbo," in *Mythologies,* trans. Jonathan Cape (New York: Hill and Wang, 1972), 56–57.

16. See Béla Balázs, *Theory of the Film: Character and Growth of a New Art* (New York: Dover Publications, 1970).

17. Claude Bragdon, *Four Dimensional Vistas* (New York: Alfred A. Knopf, 1916), 41.

18. Marcel Duchamp, *Salt Seller: The Writings of Marcel Duchamp (Marchand du Sel),* trans. Michel Sanouillet, ed. Michel Sanouillet and Elmer Peterson (New York: Oxford University Press, 1973), 94.

19. In his book *Hyperspace: A Scientific Odyssey through Parallel Universes, Time Warps, and the Tenth Dimension* (New York: Anchor Books, 1995), Michio Kaku relates this analogy from an undated short story by Nelson Bond entitled "The Monster from Nowhere." Kaku suggests that Bond was in turn influenced by Edwin Abbott's classic story *Flatland: A Romance of Many Dimensions* (London: Seely, 1884).

20. Slavoj Žižek, "Why Does the *Phallus* Appear," in *Enjoy Your Symptom: Jacques Lacan in Hollywood and Out* (New York: Chapman and Hall, 1992), 115–16.

21. Ibid., 123.

22. Henderson, 163.

Figure 7.1 *Exxon Morphing Tiger.* Digital effects produced for Exxon Oil and McCann-Erikson (Houston) by Pacific Data Images.

7

Meta-Morphing **"At the Still Point of the Turning World"** and Meta-Stasis

VIVIAN SOBCHACK

At the still point of the turning world. Neither flesh nor fleshless;
Neither from nor towards; at the still point, there the dance is,
But neither arrest nor movement. And do not call it fixity,
Where past and future are gathered. Neither movement from nor towards,
Neither ascent nor decline. Except for the point, the still point,
There would be no dance, and there is only the dance.

T. S. Eliot, "Burnt Norton"

At this particular moment in the history of representation and in our Western culture, with its particular conceptions of time and the temporal process of a human life, is there anything that has become so quickly clichéd and yet remains so surprisingly "uncanny" as the digital "morph"—especially as it has been used to transform photo-realist animations of the human body in film and television?[1] On the one hand, in a very short time, through advertising, movies, and inexpensive software that we all can buy for our home PCs, the morph as a distinct figural phenomenon has become utterly familiar, completely banal. We are intimate with the likes of such narrative shape-shifters as the T-1000 "metal-morph" in *Terminator 2: Judgment Day* (1991) or Odo, the shape-shifting constable in the television series *Star Trek: Deep Space Nine.* We have had our fill of cars turning into tigers, of human bodies going liquid and transforming into inanimate objects, of men changing into women, Caucasians into Asians or Africans or Latinos, dogs into cats, razor-thin models into busty Marilyn Monroes. On the other hand, however, there is nothing quite so continually fascinating and deeply *unheimlich* as the elasticized ease with which these digital metamorphoses effectively echo, transgress, and transform not only the "natural" spatial boundaries of the lived body but also, in the containments of their process, the presumed articulation of human animation itself. Watching a morph, at the same time I know its human "impossibility" and strangeness, I also feel myself *identifying* with it—not with its narrative figure but with its figuration of corporeal process. My own body quickens to its effortless transformations at some deep molecular level, and I recognize the morph as strangely familiar; that is, I feel "myself" in constant flux and become aware that I am never self-identical (and possibly not even self-contained). Furthermore, in its achievement of radical change in time as materially affixed to, and contained by, a concrete body or "thing" (however malleable), there is nothing quite so paradoxical as the sense the morph conveys that the ephemerality of flux and the temporality of transformation are themselves undone. Flux and transformation suddenly lose their quotidian human (if not philosophical) logic and value—not only because the morph's radical "quick-change" is fixed in a concrete and particular figure but also because its transformations are temporally reversible. What might I mean by this? To explain, I want first to consider the temporal nature of digital *morphing* as a primary mode of formal *figuration*—and then I will go on to consider the digital *morph* as a secondary discrete *figure* or "chronotopic" cohesion of narrative time and space that variously emerges to transform specific photorealist representations of the animate human body and its meanings.[2]

Transformation and Temporality

It is illuminating to contrast digital morphing with its figural cinematic counterparts: the "cut" and the "long take." (The "quick-change" of the former is associated with Sergei Eisenstein, V. I. Pudovkin, and "montage"; the more "evolutionary" transformations of the latter, with André Bazin and "mise-en-scène.") In terms of classical theories of cinema that address the medium's construction of temporal meaning and hence value, both these modes of figuration and their temporal structures stand in sharp relief to the temporal process—and meaning effects—of digital morphing. Using the cut and montage, early Soviet filmmakers demonstrated again and again how temporality and meaning are determined by the order of separate images both juxtaposed and connected by the cut. A reversal of two or more specific images will alter their temporal relations and consequently their meaning. Even when montage editing is more associative than linear in its attempts to represent change in time, even when it uses the softer "dissolve" between shots, a certain temporal necessity, a certain phenomenological sense of time's irrevocable gravity, remains as shot follows shot and the film unreels linearly before us.[3] Furthermore, use of the cut or dissolve to effect a change in time and meaning produces a *visible* (if often transparent) temporal gap or ellipsis that always points not only self-reflexively to its own internal temporal omissions but also to some external intentional agent laboring in humanly perceived "real" or mortal time to re-member human temporality's irreversible temporal process and spatial progress—that is, its entailment in our embodied experience with certain physical laws and material effects.[4]

For example, watching the series of dissolves that transform the body of poor Larry Talbot into a werewolf in *The Wolf Man* (1941), although we may be thrilled by the horror film's transgression of the physical laws according to which we as human spectators live, we are also aware that the metamorphosis represented before us is visibly marked by temporal gaps—during which, exempt from our sight and enduring hours of makeup and stop-and-go filming, the physical body of the actor Lon Chaney Jr. was being made more or less lupine and hirsute. (The record indicates that applications of the rubber snout, fangs, claws, and yak hair took nearly five hours to complete and that shooting the transformation scenes through stop motion substitution took as long as twenty hours and involved bracing the actor's head to keep it in the same position while makeup artist Jack Pierce built up the illusion from shot to shot.)[5] Thus although Larry's lycanthropic transformation in the narrative is ostensibly reversible (that is, he changes back to human form in the morning), one could argue that as formal figuration, the dissolves from shot to shot that constitute the film's temporality are not. Indeed, this very difference between the deep temporal structure of the mode of figuration and the surface temporality of the narrative resonates in a way that grounds the poignancy of Larry's condition in time: that is, although his various transformations into a wolf man and back to human form are visibly reversible to the human eye, his horror film malady is itself continuous and irreversible—and only death will cure him of it.

In the case of the long take, Bazin and other post–World War II cineasts who favored "realism" privileged a film style that avoided montage and thus allowed "natural" transformations to "emerge" in what seemed homologous to humanly experienced change in "real time" as well as contiguous space. Keeping animate subjects in the camera's gaze in a single shot of lengthy duration constituted the existential meaning of transformation and change as a visibly uninterrupted process of "becoming"—the value of such transformation conferred by the physical labor and teleological progression entailed in mortal temporality as it is writ upon, and understood by, the Western human body. Relating our understanding of the meaning and value of temporality on the screen to the spectator's own embodied perception of lived time and transformation, Bazin privileged the long take over montage even when it was used in the service of photo-realist fantasy.[6] Thus it is likely that he would have far preferred the transformation sequence in *Dr. Jekyll and Mr. Hyde* (1941) to that in *The Wolf Man*—for Spencer Tracy "becomes" Dr. Jekyll in front of the camera and our eyes, not through an elided process involving unseen applications of makeup and prosthetic devices but through the visible labor and duration of the actor's performance in its uninterrupted metamorphosis. Even when transformations of character and action in the long take evolve in less flamboyantly dramatic fashion or are further complicated by camera movement and by the rich simultaneity of spatial relations that emerges through correlative use of the distancing "long shot," the temporal meaning of what has come to be identified as "mise-en-scène" editorial practice is essentially linear and progressive—and valued precisely for its nonreversibility.

Thus, whatever the differences between montage and mise-en-scène, both cinematic modes of representing change in time produce temporal meaning and the elements that constitute it as formally nonreversible. Furthermore, the transformations in time effected by both visibly present themselves as the consequence of temporal and physical *labor*—either, in the case of montage, of the filmmaker splicing away to make symbolic meaning of a particular temporal order or, in the case of the long take, of characters and objects in the phenomenal world encountering and overcoming obstacles as they endure and change according to the demands of what is perceived by a physically embodied and mortal spectator as "real" existential time.

In terms of photo-realist cinema, then, perhaps the only figuration of temporality we experience as truly reversible is the reversed motion of the film itself—accomplished either through shooting with the film reversed in the camera, through reversals effected in optical printing, or through the reversal of the projector during exhibition. Indeed, in early cinema, "trick films," their film loaded backward during shooting and then projected normally, foregrounded such physical and temporal impossibilities as someone leaping upward to incredibly great heights or broken fragments of china reforming themselves into once again coherent crockery. Projectionists would also often amaze their audiences by literally cranking the film backward. Even today there is an uncanny, if also often playful, aspect to watching the figure of reverse motion on the screen (achieved now primarily through optical printing). The beginning of *The Crimson Pirate* (1952) has Burt Lancaster swinging acrobatically by rope from right to left across the screen to directly address the audience, telling us: "Believe only what you see"—at which point, the film ostensibly reversing itself like a pendulum, he swings impossibly backward and upward from left to right, lands on the ship's yardarm from which he started, and amends his initial directive with a grin: "No. Believe half of what you see." Even more *unheimlich,* however, is such figural reverse motion when it is put to the service of a narrative or thematic emphasis on the essential irreversibility of human temporality and physical animation, and on mortality. In these instances, metaphysics overtakes mere trickery. Thus the haunting power of reverse (and slow) motion as it brings "back to life" the dead young poet Cegeste in Jean Cocteau's *Orpheus* (1949), or as it confers an aura of death to Standish Lawder's *Necrology* (1970), an aptly titled one-shot film photographed completely in reverse to make visible a rush hour crowd of people standing on the down escalators leading into New York's Grand Central Station—all of whom stare blankly ahead as they ascend backward and upward out of the frame.[7] In photo-realist cinema, such visible figures that reverse both human temporality and the grounds of cinematic temporal figuration are relatively few and far between. One might argue that this rarity is merely the result of convention. However, in the context of the discussion here, it is worth considering that the infrequent figuration and figure of reverse motion is the result of a tension in the medium between, on the one hand, the cinema's linear temporal grounding as well as its photo-realist reference to a world governed by the physics we humans live and, on the other, the most extreme attempt completely to reverse this temporal teleology and that of the humans subjected to it. Thus it is telling that reverse motion is used most when it is used most invisibly—and most literally—to reduce the mortal peril of human bodies performing such dangerous stunts as car crashes.

In contrast to its occasional and uncanny appearance in the cinema, temporal reversibility in the realm of the digital seems not only common but also strangely "natural." Indeed, insofar as our sense of "naturalness" is tied culturally, historically, and philosophically to ontological notions of the seamlessly "sheer" being of Being, to Being's pretheoretical transparency and effortlessness, one could argue that to us humans, who cannot so easily bend our substantial form to our will and must struggle in our quotidian lives to realize our being, it is precisely this very sense of "natural" effortlessness that makes morphing as figuration and the figure of the morph so uncanny.[8] In particular, as a visible figure, the morph confronts us with a representation of Being that is intellectually familiar yet experientially uncanny. It calls to the part of us that escapes our perceived sense of our "selves" and partakes in the flux and ceaseless becomings of Being—that is, our bodies at the cellular level ceaselessly forming and reforming and not "ourselves" at all. Thus the morph is not merely a visible representation of quick and easy transformations of matter in time and space: it is always also an oxymoron, a paradox, a *metaphysical object.*

Unlike its cinematic predecessors, digital morphing as transformative figuration and specific narratological figure of transformation both constructs meaning as reversible and visibly represents transformation as a consequence more of will than of external or internal temporal and physical labor. Indeed, counter to the temporal evolution and material encounters of the long take privileged by a modern humanist such as Bazin, we could say that the digital morph as narrative figure takes all the *bildung* out of the *bildungsroman* (an issue to which I shall later return).[9] As figuration, morphing presents no reflexive temporal "gap" or "ellipsis" like the cut or dissolve, and as a hermetic and self-absorbed figure evolving without interruption in the mise-en-scène of a cinematic narrative, it finds there no inherent obstacle to the sea change of its own becoming. In this regard, "destroying" the morph as narrative figure presents major logistical—and logical—spatiotemporal problems: How to stop its shape-shifting? How to fix its dispersals? Indeed, "killing" a morph is an impossible business: insofar as the verb is humanly meaningful, its use seems a curiously inappropriate way to speak about the figure's "termination," since "killing" entails notions not only of bodily animation but also of bodily coherence, human temporality, and mortality.[10] In *Terminator 2,* the T-1000 is riddled with shotgun blasts, but the sides of his wounds immediately "run" together to fill the bullet hole; or he is frozen ("fixed") with liquid nitrogen and then shattered only to have the pieces liquefy and reunite into a whole; he is "finally" dispatched (and dispersed) by being thrown into a vat of molten steel, though the finality is merely narrative, not substantial. Thus the uncanniness of the morph is that its conceptual coherence as a figure of transformation is dependent on its literal incoherence as a "fixed" figure. Usually, its "end" is, in fact, a return to its "beginnings" in an amorphous soup of shifting matter.

Indeed, with its end in its beginning (and vice versa), digital morphing transforms the very grounds of a cinematic temporality tethered to—if not completely bound by—gravity: gravity as a value of photographic indexicality to a spatial and material world, to the visibility of particular human and representational labors marked by change in space and time, and to human mortality. In *The Wolf Man,* Larry's cinematic transformation from man to wolf to man thus matters and has a particular human temporal value that is not reversible in either meaning or substance even if it is in form—and the same is true of *Dr. Jekyll and Mr. Hyde.* In *Terminator 2,* however, the digital transformation of the "liquid metal" T-1000 from man to tiled linoleum floor to man is temporally reversible, its meaning having nothing to do with human temporality—or matter. That is, the "man" who appears as the T-1000 was never a man *originally,* and thus his transformation into the tiled floor has no hierarchical value in either form or substance. Paradoxically, then, morphing and the morph deflate in humanly meaningful temporal value proportionate to their inflated spatial display of material transformation as both seamlessly reversible and effortless. This, of course, is a phenomenological effect we experience regularly when we engage the playful cartoon physics of animated films. However, when such an impossible physics irrevocably alters verisimilar forms of photographic representation and human action to digitally warp and punctuate (if not completely puncture) the temporal gravity and spatial grounds of cinematic realism, the effects of temporal evacuation and spatial shiftiness seem less playful than *unheimlich.*

In some instances, the effects of digital morphing as a mode of figuration are relatively transparent, and as with the verisimilar yet talking animals in a film such as *Babe* (1995), one looks through them while focused primarily on narrative elements. In other instances, however, morphing's transformation of, and challenge to, the very grounds of cinematic indexicality and representational labor are foregrounded to become allegorical figures of this very problematic. A film such as *Dark City* (1998) serves as paradigmatic. The morphological ground of the film's cinematic space and time not only is transformed at a deep structural level by digital effects but also is narratively allegorized and visibly figured. Located in some unknown metropolis, the science fiction film's "Strangers"—in a communal act of will and wish—effectively "stop time" at the stroke of each midnight, literally and visibly warping, expanding, shrinking, and shifting buildings and streets to materially and spatially transform an entire cityscape that is, nonetheless, fixed and bounded as a finite world. Correlatively, in secret experiments, the Strangers also literally re-member the temporality and memory of the human inhabitants who live within the shifting city's quite finite boundaries. Just as the Strangers radically transform the space of the city and the temporality of its inhabitants, so does morphing, as figuration, radically transform the spatial and temporal grounding of a photo-realist cinema that up until now has been indexically related to human physical existence as it is daily experienced in space and time.

Assimilating Difference

As a discrete figure that punctuates photo-realist narrative and physics, the morph hyperbolically dramatizes and, more significantly, *comprehends* and *contains* not only quick-change in time but also radical difference in form and matter. Indeed, quick-change in time and radical difference in form and matter are the morph's raison d'être—and yet their very hyperbolization by the morph is also a reduction in which transformation as a raison d'être empties quick-change of its temporal specificity and existential value and comprehends radical difference within the bounds of an ultimately self-similar and unitary figure. Thus, it could be argued that as a paradoxically bounded and fixed figure of transformation, the morph operates to superficially simulate change as, on a deeper level, it assimilates not merely "difference" but also "otherness."

Here I would point not only to the blatant demonstration of this argument in the well-known and easy morphological assimilation of different racial/ethnic "happy faces" in Michael Jackson's video *Black or White* (1991) but also to Woody Allen's *Zelig* (1983)—an earlier and much more reflexive cinematic counter-text that, although it does not accomplish its morphing through current digital figuration, nonetheless explores in some detail the strange paradoxes of self-identity, bodily transformation, and the assimilation of ethnic and racial difference. Leonard Zelig is a human chameleon, so desperate to "fit in," to be "the same" as "the other," that he assumes the physical characteristics and ethnicities (as well as the languages and occupations) of those around him to become literally black, Italian, Chinese, French, Irish, and Greek. At the same time, however, his assimilation of difference is parodic, thwarting its own goal, challenging the very possibility of "becoming other" at the very moment it asserts "becoming other" as inherent to the human condition. Caught up in a dynamic of continual transformation and always provisional assimilation, "Zelig's own existence," as the voice-over narrator tells us, "is a nonexistence"; he is "a cipher, a nonperson"; he is not "self-possessed." *Zelig* calls explicit attention to the paradoxically both real and fictive status of self-possession and (ethnic and racial) identity, interrogating the conventional nature of their "fixity" and emphasizing their very existence as always provisional and of the moment. Indeed, the instability and evanescence of Zelig's self-sameness paradoxically points not only to his pathology but also to every human being's very normal construction and revision of himself or herself as always continually self-different. Thus the newspaper headline in *Zelig* that announces, "Human Who Transforms Self Discovered," refers not only to Zelig but also to all of us—and like the figure of the digital morph, it denies while it affirms traditional notions of both self-identity and the potential for either becoming the difference that is "the other" or assimilating the other's difference to our own "self-sameness."[11]

In this regard, it is particularly telling that the digital morph as a figure of transformation tends to transform itself between two poles that are clichés of opposition. (One sees this most clearly in both morphing tutorials and commercial advertising.) That is, the morph's primary mode is to assert not only sameness *across* difference but also the very sameness *of* difference. While often representing cultural binaries at its static end points (i.e., gender, race, species, and subject-object oppositions), the process of the morph attempts to erase this binarism in the homogeneous, seamless, and effortless movement of transformation and implied reversibility. (And here I would argue that whether one actually *sees* the reversal in narrative use is irrelevant to our phenomenological knowledge of its possibility.) Indeed, in the process of the morph, not only "difference" but also "otherness" are effaced at the very moment of their foregrounding—a condition best dramatized by Jackson's *Black or White.* As Jean Baudrillard might describe it: "A determinate negation of the subject no longer exists: all that remains is a lack of determinacy as to the position of the subject and the position of the other. Abandoned to this indeterminacy, the subject is neither the one nor the other—he is merely the Same."[12] In the name of an ill-conceived multiculturalism, the music video collapses both difference and otherness into self-sameness as we watch a range of human faces distinctly marked by their difference and otherness morph one into the other in a reversible chain not of resemblance but of smiling similitude.

Michel Foucault provides a useful distinction between the two quite different logics that inform *resemblance*, on the one hand, and *similitude* on the other. The stuff of metaphor (and the cinematic cut and dissolve), relations of resemblance may assert a form of sameness, but they "demonstrate and speak across difference" and are temporally (and culturally) hierarchical, requiring the "subordination" of one term to another term that provides the "original" model for the comparison.[13] That is, something is *like* something *else*. It is here that temporal ordering confers cultural value and significant meaning. If the digital morph were not essentially reversible, it would matter significantly, rather than trivially, that a woman in the tutorial was turned into a man or a lynx, an African American into a Caucasian. One term of the transformation would be hierarchically dominant; what resembled what would be semantically important in its specific temporal order.

Relations of similitude (and the morph), however, may foreground and assert difference, but indeed, they demonstrate and speak across sameness. They are nonhierarchical and reversible. As Foucault explains:

> Resemblance has a "model," an original element that orders and hierarchizes. . . . Resemblance presupposes a primary reference that prescribes and classes. The similar develops in series that have neither beginning nor end, that can be followed in one direction as easily as in another, that obey no hierarchy, but propagate themselves from small differences among small differences. Resemblance serves representation, which rules over it; similitude serves repetition, which ranges across it. Resemblance predicates itself upon a model it must return to and reveal; similitude circulates the simulacrum as an indefinite and reversible relation of the similar to the similar.[14]

It is apposite that Foucault makes these distinctions between resemblance and similitude in *This Is Not a Pipe*, his meditation on the work of René Magritte. Indeed, a good many of Magritte's paintings reveal to us the uncanny and paradoxical moment in which analogy without value and difference without hierarchical succession transform transformation itself into a static essence. (*The Explanation* [1952], in which a wine bottle and a carrot combine properties into an impossible object, *The Empire of Light* [1954], in which the binaries of day and night both announce and lose their distinction, and *The Exception* [1963], in which a cigar and a fish merge seamlessly one into the other, are only three such examples.)

Unlike most other signs of transformation, and formally akin to these

Magritte paintings that freeze transformation into permanent flux as they simultaneously assert mutability and difference in time, the morph is *palindromic.* That is, it can be read forward and backward without a change in meaning. It is no mere coincidence, then, that the "shape-shifter" of *Star Trek: Deep Space Nine* is named Odo.[15] Indeed, not only does his name read reversibly, but so do his nature and function. On the one side, in his "fixed" or "arrested" humanoid state, Odo—the constable—spends his time "arresting" those who transgress the boundaries of law and order. On the other side, however, Odo himself must take a "rest" every sixteen hours, relaxing his humanoid shape into a transgressively "unfixed" and gelatinous state whose formlessness is "contained" by a bucket. Odo "at rest" in his bucket as "flux"—arrested, as it were—is both a narrativization and gloss on the essence of the morph. He is pure change as a reversible form without meaningful shape: that is, form cleansed of causation, robbed of specificity and reason, drained of temporal value. Transformation and all its consequences are assimilated to the notion of a palindromic "eternal return" (to the bucket?), to the rhythm of recurrence, to a totalizing circularity in which everything is connected, subsumed, and become equal.[16] This, indeed, is precisely articulated in the series' narrative as the ideal life on Odo's distant home planet; his "rest" as "dissolution" in a meager bucket on an alien space station is dramatized on his home world as a literal ocean of communal and unchanging sea change, a formless dispersion and flux of self-identical being. As one of his species tells him, with no hint of irony: "You are a changeling. You are timeless."

Haunted by Similitude

The temporal reversibility and palindromic quality of the morph, the nonhierarchical similitude of its elements, may seem in their leveling of difference somehow "democratic." Yet at what cost? The value of time collapses, and difference is accumulated not as a whole constituted from discrete elements but rather as a subsumption to the sameness of self-identity. It is instructive to contrast once again the phenomenological—and material—effects of the cinematic cut and the digital morph. Here a montage sequence in the opening dance audition scene of *All That Jazz* (1979) sits in interesting relation to Jackson's use of digital morphing in *Black or White* to create one unified "body" from human beings disparate in gender, race, and ethnicity.

The montage in *All That Jazz* is dazzling in its precise cutting of a gender and racial mix of dancers' bodies together to constitute the action of a single pirouette. However, while the action in space is continuous, the bodies in time are not; we are still aware of their discretion and difference. The effect of the cinematic montage is not to undo time or rob it of value but to compress it. What is conflated through the montage of the perfectly cut pirouette, then, is not the discretion and difference of the dancer's *bodies* in their moment of audition but their self-similar *aspirations*. This is, indeed, a sequence in which "resemblance serves representation." The digital morphing of disparate bodies in *Black or White*, however, achieves rather different effects. The process that unifies these bodies is not merely analogous to cutting them together on an action—nor is it merely an effect of the musical continuity of the song they all sing. Seamlessly warping and morphing one into the other, these racially and ethnically "different" singing heads enjoy no discretion: each is never "itself" but rather a mutable permutation of a single self-similarity, a series of "small differences among small differences" that reduce their value to one.[17] Thus the morphing sequence develops with neither a significant beginning nor an end, and as Foucault suggests, it "can be followed in one direction as easily as in another." Temporality here is not compressed; it is conflated to the present, where difference merely haunts the same. In *Black or White*, then, "similitude circulates the simulacrum as an indefinite and reversible relation of the similar to the similar." Jean Baudrillard, in a surprisingly cautionary essay on "cloning" called "The Hell of the Same," is apposite here. Indeed, he might well have been making a distinction between the resemblance effected by the cinematic cut and the similitude effected by the digital morph when he writes: "Division has been replaced by mere propagation." And he might well have been discussing the difference between the analogous sequences of bodily transformation in *All That Jazz* and *Black or White* when he continues: "And whereas the other may always conceal a second other, the Same never conceals anything but itself. This is our clone-ideal today: a subject purged of the other, deprived of its divided character and doomed to self-metastasis, to pure repetition."[18]

This recognition that what seemed like division is merely propagation, that the Same never conceals anything but itself, is dramatized through the digital morphing of a generation of family photographs and home movies in Daniel Reeves's extraordinary—and aptly titled—autobiographical video *Obsessive Becoming* (1995).[19] This is a work that presents the dark side of *Black or White*'s self-identical "happy face" to explore and exorcise the "doom" of "self-metastasis" in both personal and public history. *Obsessive Becoming* presents a morphological genealogy of child abuse and wanton violence that haunts (re)generations of both a single family and the larger world. Although the warping and morphing of family members in and across their generations constitute the ever-shifting and seismic ground of the video, the movement is indeed meta-static: at once in constant transformation from one point to another and yet also essentially fixed in time and space at what T. S. Eliot has called "the still point of the turning world."[20] Describing a sustained sequence in which "portraits of Reeves' ancestors are morphed, generation to generation, revealing the persistence of a bloodline and a cherished history" (however abusive), Steve Seid points out: "Along with keen hand tintings and other painterly treatments, the hi-tech effect of morphing finally becomes not so much graphical pizzazz as a visual recognition of our seamless linkage to the past. . . . *Obsessive Becoming* is an attempt to acknowledge . . . failings and then, hope against hope, thwart the age-long cycle of cruelty."[21] Indeed, ripe with rage and regret on the one hand and love and reconciliation on the other, Reeves's voice-over narration invokes a timeless litany of both family dispersion and eternal return. "These faces are windows into the house of no escape. They open into one another and sing with the billions of doves. They are no different than you—the face is the same," he tells himself (and us) at one point; and at another, "These faces keep leaving and these faces continue to return and have never left"; and at yet another, "This fire is all of one family."

What is extraordinary about *Obsessive Becoming* is that it uses morphing and the morph to assert the paradox of human existence as at once both timeless and historical. Although haunted by the "hell of the same," Reeves grounds the morphological flux of timeless self-similarity in the fixed specificity and the value of small details, of singular—hence historical—lives. Again and again, and against the shifting, warping ground of generational history, Reeves stresses that "the morph stops here"—that the ebb and flow of transformation, while (re)current, becomes itself the object of reflection, emotionally arrested by quite specific bodies that give it particular and present autobiographical meaning. Thus, while asserting similitude, through an affecting representational labor of love that gives human (not computergraphic) emphasis and value to the lives and times he weaves together, Reeves ultimately discovers resemblance—and the possibility of difference, of breaking repetitive and self-similar cycles of violence and abuse. This is at far remove from the metastatic and cancerous consumption of difference by difference in *Black or White*. Unlike Reeves's work,

Jackson's video is hardly haunted by similitude: rather, it flaunts similitude in the very guise of human difference.

Being in Labor

I have just described the morphological achievement of *Obsessive Becoming* in its uneasy but hopeful transformation of similitude to resemblance, of temporal reversibility to human history, as a "labor of love." But the achievement of human history seems always also to entail a "love of labor." Indeed, throughout this essay, I have contrasted the morph's effortless animations, its imperviousness to obstacles, its very ease of becoming, with the labor associated with human beings' temporal emergence and self-realization.

In relation to the process of human lives, the morph's effortless and elastic ease at "realizing" itself is deeply uncanny: strangely familiar insofar as much of our physical existence is something transparently operational and continuous to us, unfamiliarly strange insofar as the realization of most of our conscious acts involves some degree of hesitation, difficulty, and effort. Indeed, within the historical transformations of the narrative tales we, in Western culture, tell ourselves about human becoming and self-realization, those that are about the movement and formation of what we value as "self-identity" or "character" are measured in chronotopic terms of obstacles, ruptures in action, frustrations of desire, difficulty and effortfulness in "being all that one can be." Thus, in human terms, being is narrativized in time and space as the culmination of a laborious business—first in the effortful process of one's birth and then in the effortful progress of one's life. Hence the uncanniness of metamorphosis as it is figured generally throughout Western literature, marked as it is by sudden appearance, quick physical change in space, and effortless transformation in time. Hence the "realism" of the bildungsroman (the novel of education) and other novels of "human emergence" in which physical transformation and action in space and the creation of human character do not just happen *in* time but also *take* time. Is it any wonder, then, that in those photo-realist narratives in which the figure of the morph takes on human form, the digital morph—as a "character"—confounds us?

I want to return to the palindromic, shape-shifting "character" of Odo, and to "The Begotten," an episode of *Star Trek: Deep Space Nine* whose narrative purposely—and (computer)graphically—plays with and confuses the entailment of labor and being in relation to both humans and morphs.[22] The episode follows and entwines two narratives—a primary one in which Odo finds an "infant" changeling and, with his own mentor Dr. Mora, attempts to teach it how to morph and how to hold a shape, and a secondary one in which pregnant Major Kira (a female and humanoid Bajoran crew member on the station) goes through labor and a Bajoran relaxation ritual to give birth to a human baby boy.[23] The irony of the episode is that the infant changeling, in an initial condition of relaxed formlessness, has to learn to labor in order to "hold" a specific shape, whereas the pregnant Kira, in labor, has to learn to relax in order to give birth to her baby.

In the Odo narrative, the effort and difficulty and even pain of "becoming" are seen as essential to its process. Initially Odo does not want to "prod" and "poke" and "shock" the infant changeling into the intentional movement of shape-shifting, thus replicating the painful experience Odo had at the hands of Dr. Mora (who wasn't even sure Odo's gelatinous mass was a sentient life-form). "You enjoyed watching me suffer," Odo says to Mora, who replies: "If it wasn't for me, you'd still be sitting on a shelf somewhere in a beaker labeled 'Unknown Sample.'" Refusing unpleasant provocation as a method, Odo gently introduces the semiliquid changeling to a variety of forms, pouring it into beakers shaped like a sphere, a cube, a pyramid, in hope of getting it to mimic their shapes on its own—but to no avail. The changeling remains unresponsive, so Odo tries coaxing:

> I understand how you prefer to remain shapeless. Believe me, I remember how relaxing it could be, but you have to learn to take other forms. That's what changelings do. It can be immensely rewarding. I remember the first time Dr. Mora here coerced me into taking the shape of a cube with one of his electrostatic gadgets. Once I did it and he turned the infernal thing off, I was perfectly content to stay a cube for hours. It was fascinating—all those right angles.

Reluctantly, Odo agrees to try a bit of Dr. Mora's shock treatment on the unresponsive and unformed form—but he doesn't set the voltage high enough to provoke any movement, and Dr. Mora chides him: "Odo, without discomfort the changeling will be perfectly comfortable to remain in its gelatinous state. It'll just lie there—never realizing it has the ability to mimic other forms, never living up to its potential." Mora might well be paraphrasing Sigmund Freud in *Beyond the Pleasure Principle:*

> *Instinct is an urge inherent in organic life to restore an earlier state of things* which the living entity has been obliged to abandon under the pressure of

> external disturbing forces; that is, it is a kind of organic elasticity, or, to put it another way, the expression of the inertia inherent in organic life.[24]

Of course, the changeling does eventually respond in proper morphological manner—not only taking on geometrical shape but also roughly casting itself to mimic Odo's humanoid form. But Odo's parental labor of love and the changeling's newfound love of labor cannot continue, for as one character aptly tells Odo when he announces he's finally happy, "If you're happy, there is something very wrong in the world. The center cannot hold." At least not for a shape-shifter. And so the once-sick infant changeling gets ill again and "dies"—literally dispersing itself into Odo's "body" and reenergizing his own power of metamorphosis as it does so.

The secondary narrative of Kira's labor is comic—and slight—in comparison, but it overtly reverses the problematic of the infant changeling's emergence as an effect of pain and effortfulness. Its main emphasis is on the opposing difficulty Kira has in achieving a sufficient state of relaxation to go through the quick and painless labor of actually delivering the baby. In what amounts to a hyperbolization (and parody) of Lamaze "natural" birthing methods, a Bajoran priestess shakes a rattle, and the baby's two biological parents clumsily attempt an accompanying rhythm on other ritual instruments while Kira tries to relax. Her "efforts" to do so are variously confounded, not least by the bumbling and argumentative men attending her. "I'm *trying* to have a baby," she barks at them in exasperation. Finally, the baby is born—but to words that perversely reverse Odo's instructions to the emergent figure of the changeling infant: "That's it, relax. Let it come."

This particular episode of *Deep Space Nine* is even more complex than I suggest in its exchanges and reversals between the chronotopic development of humans and morphs and between the ease and labor necessary to animate being. It is, however, despite its transgressive play with transformation and becoming, a fairly traditional bildungsroman. That is, the episode as a whole is akin to what Mikhail Bakhtin calls the "novel of education." In it, while educating the infant changeling, Odo is himself educated: he learns to accept the "prodding" and "poking" that is necessary for being to take on a determinate and meaningful shape, and he learns to value the labor that gives being "character." Thus despite its emphasis on "becoming" and its use of digital morphing, "The Begotten" begets a notion of human emergence that is historically anterior both to the imagination of its own narrative science fictional future and to the present imagination of viewers watching the episode in the contemporaneous digital moment.

History and Metamorphosis

In his extraordinary gloss on the spatiotemporal history of novelistic structures, "Forms of Time and Chronotope in the Novel," Bakhtin takes up early chronotopic transformations of the image of transformation itself, particularly within the compass of what he identifies as the literature of "Greek adventure-time."

> Metamorphosis or transformation is a mythological sheath for the idea of development—but one that unfolds not so much in a straight line as spasmodically, a line with "knots" in it, one that therefore constitutes a distinctive type of *temporal sequence.* The makeup of this idea is extraordinarily complex, which is why the types of temporal sequences that develop out of it are extremely varied.[25]

Bakhtin notes, for example, that in the work of Hesiod, the word "metamorphosis" is used in relation to cosmological forces and to genealogical series of ages and generations. It does not carry with it "the specific sense of a miraculous, instantaneous transformation of one being into another" as it does in Ovid's *Metamorphoses*—nor is it yet attached to the transformation of "individual, isolated beings."[26] These detachments of metamorphosis from the cosmic and genealogical appear only later in the Roman and Hellenistic era. Evoking the Roman image of Caesar metamorphosed into a star, Bakhtin writes of these new figurations of transformation—but his description seems also relevant to certain figurations of the morph today: the T-1000 morphing into a tile floor or the meta-static but discretely unitary face of *Black or White:*

> They are metamorphoses only in the narrower sense of the word, changes that are deployed in a series lacking any internal unity. Each such metamorphosis suffices unto itself and constitutes in itself a closed, poetic whole. The mythological sheath of metamorphosis is no longer able to unite those temporal sequences that are major and essential. Time breaks down into isolated, self-sufficient temporal segments that mechanically arrange themselves into no more than single sequences.[27]

Bakhtin charts further changes in this premodern morphological imagination. In Apuleius's *Golden Ass,* Bakhtin sees a novel "personalization" of metamorphosis that now has little meaning as a "natural" cosmological force but takes on a "quite openly magical nature." Here, "metamorphosis has become a vehicle for conceptualizing and portraying personal, individual fate, a fate cut off from the cosmic and the historical whole," yet energetic enough "to comprehend the *entire life-long destiny of a man,* at all its critical *turning points.*"[28] These are figural changes that Bakhtin sees as key to the emergence of what will eventually become the modern, "realistic" image of human transformation—although he points out that this "crisis portrayal" of *"how an individual becomes other than what he was"* is not yet evolutionary, not yet biographical.[29] Nonetheless, for Bakhtin, Apuleius *humanizes* Greek adventure-time and grounds his protagonist's transformation into an ass in a temporal suffering and physical labor that is connected to individual responsibility and ultimately leads to individual enlightenment. Noting this critical move away from the purely mythological notion of metamorphosis and toward human biographical evolution, Bakhtin writes of *The Golden Ass:*

> Here time is not merely technical, not a mere distribution of days, hours, moments that are reversible, transposable, unlimited internally, along a straight line; here the temporal sequence is an integrated and *irreversible whole.* And as a consequence, the abstractness so characteristic of Greek adventure-time falls away.[30]

Thus begins the "knotted" historical movement to the bildungsroman and its precedent and antecedent novelistic forms: to Odo, whose abstract palindromic reversibility in both name and substance is arrested in the figure of the space station's constable and assumes the whole "character" of an irreversible human temporality not only marked—but made—by labor, suffering, responsibility, and care.

My aim here is not to wax nostalgic. Indeed, it is to look forward and to understand the present uncanny presence of the morph as a *digital* figure of human emergence, one not substantially "in-formed" by a human actor playing at being a morph in the same way the morph he becomes plays at being a character. After all, *Star Trek,* in its many manifestations, has always been first and foremost a *humanist* text—and so it is not very surprising that it would contain its digital flux within the old tin bucket of the eighteenth-century bildungsroman. If we are to think historically and dialectically about the morph, however, we cannot contain it in such a traditional interpretation of emergence. On the one hand, referencing the T-1000 and the *Black or White* "happy face" to the reversibility and abstraction of Greek adventure-time or Odo to the bildungsroman may be heuristic, but on the other, it also suggests that the more things change, the more they stay the same. Thus we need to remember that the T-1000, Jackson's multicultural simulacrum, and even Odo are novel figures of transformaton, historically different from what has ever been imaged and seen before—which is also to say that the more things stay the same, the more they change.

In "The *Bildungsroman* and Its Significance in the History of Realism (Toward a Historical Typology of the Novel)," Bakhtin continues the consideration of historical changes in the literature of human metamorphosis begun in "Forms of Time and Chronotope in the Novel." Cautioning that "no specific subcategory upholds any given principle in pure form," he nonetheless lays out a historical array of chronotopic prevalences of novelistic forms that variously deal with the "image of *man in the process of becoming*."[31] These forms include but do not culminate in the bildungsroman. Each advances a more temporally complex hero and moves increasingly away from "adventure-time" toward a "realistic" vision of human temporality as "limited, unrepeatable, and irreversible."[32] And some few become "novels of emergence" and significantly *historical*. That is, in them, individual human emergence is inseparably linked to the emergence of the world. This is the form of novel Bakhtin most privileges. "In it," he tells us, "man's emergence is accomplished in real historical time, with all of its necessity, its fullness, its future, and its profoundly chronotopic nature."[33] Oddly, however (or so it first seems), along with expected examples such as *Wilhelm Meister*, Bakhtin holds up Rabelais's hyperbolic and expansive *Gargantua and Pantagruel* as a model. Here,

> human emergence is of a different nature. It is no longer man's own private affair. He emerges *along with the world* and he reflects the historical emergence of the world itself. He is no longer with an epoch, but on the border between two epochs, at the transition point from one to the other. This transition is accomplished in him and through him. He is forced to become a new, unprecedented type of human being. What is happening here is precisely the emergence of a new man. The organizing force held by the future is therefore extremely great here—and this is not, of course, the private biographical future, but the historical future. It is as though the very *foundations* of the world are changing, and man must change along with them. Understandably, in such a novel of emergence, problems of reality and man's potential, problems of freedom and necessity, and the problem of creative initiative rise to their full height. The image of

> the emerging man begins to surmount its private nature (within certain limits, of course) and enters into a completely new *spatial* sphere of historical existence. Such is the last, realistic type of novel of emergence.[34]

Bakhtin both presents us with and glosses a conundrum relevant not only to his own chronotopic history of representation but also to ours in the present moment of human emergence. The conundrum is that the novel in which human being emerges as fully historical is a novel that is unrealistic, fantastic, incredible, impossibly excessive and expansive. In *Gargantua and Pantagruel,* transformation is figured quite literally as the norm—heroes and world each adjusting to the other in what seems an exaggerated and impossible expression of their chronotopic and morphological interdependence. Human bodies leak, shed, shit, become flatulent, expand, consume and cannibalize the world. The world, too, seems always to change its shape in response. For Bakhtin, then, this patently "unrealistic" novel is a new form of realism that finds a new and appropriate chronotope with which to figure the emergence of a historically novel form of human being. Again, what must strike us here is how extraordinarily well Bakhtin's description of *Gargantua and Pantagruel* also describes both the figurations and the figure of the digital morph. Reading it, I cannot help thinking not only of the emergence of a "new unprecedented type of man" like the T-1000 but also of the emergence of a new world—both real and science fictional—whose very "foundations" are changing, as in *Dark City.* Doesn't the T-1000 quite literally emerge "along with the world" and reflect "the historical emergence of the world itself"? Isn't he, the new man, "no longer with an epoch, but on the border between two epochs, at the transition point from one to the other"? This transition is "accomplished in him and through him," and thus we see the historical transubstantiation of the substratum of being from carbon to silicon. Furthermore, this image of the emerging man begins "to surmount its private nature" and enters "into a completely new *spatial* sphere of historical existence." So, it seems, the more things change, the more they stay the same: *Gargantua and Pantagruel, Terminator 2: Judgment Day.*

And yet if we take seriously Bakhtin's notion of the chronotope as an "optic" for looking at the historical relations between culture and spatiotemporal figuration, we must also acknowledge that we are rereading his description through the lens of the present—much like the critic in Borges's "Pierre Menard, Author of *Don Quixote*" who rereads the exact "same" passage written first by Cervantes and then by the more modern

Pierre Menard and notes their radical difference.[35] Meanings morph in time, and so do possibilities for representation. The more things stay the same, the more they change. Thus the "new man" in our current realist fictions emerges not through an exaggerated emphasis on the human process of living as a body, enduring time, belching and farting existence and the world into meaningful being. He comes to us (and, at the moment, he is definitely a "he") as an ascetic expansion of computation, a slight and light body occupying a different world whose gravity and physics are in question, as is perhaps also the very sense of what Bakhtin would think of as history. Yet Bakhtin himself understood this: "In such a novel of emergence, problems of reality and man's potential, problems of freedom and necessity, and the problem of creative initiative rise to their full height." Thus not only has the novel of emergence now transformed both the world and representation previous to it, but—in ways that Bakhtin could not have envisioned (except in its general form)—"the image of the emerging man begins to surmount its private nature (within certain limits, of course) and enters into a completely new *spatial* sphere of historical existence. Such is the last, realistic type of novel of emergence."

Bakhtin's "certain limits" are not our own. The morphological "new man" is not necessarily the "hero" of our world—at least as we once imagined it as real and still play out its past forms in the present of our fictions. Indeed, as this "new man," the figure of the morph (particularly as it appears in the most tired of our current moving images) both confirms the existence of and does battle with the old and stable "hero" figure, who is able only to disperse its transformative power, but not to destroy its future. This is what Bakhtin understood in choosing *Gargantua and Pantagruel* as the realistic novel of its time: realism itself is always taking on new temporal and spatial forms.

Ending at the Beginning

My argument has constructed the digital morph (in ways that it has been constructed structurally and materially) as dialectical and dialogical—as a paradoxically metaphysical yet concretely visible object of desire: some thing that is a nonthing of lived human history and ideal Kantian purity. That is, it represents and enacts human transformation, metamorphoses, mutability outside of human time, labor, struggle, and power, seemingly transcending structures of hierarchy and succession. Indeed, its uncanny effect emerges precisely—and historically—from its articulation of Western metaphysics in both its specific form of figuration and its visible and concrete form as an image. The morph is at once science fictional and realistic. As an image, it is understood both as "something through which the mind contemplates the flux of becoming" and as "something discerned inside that flux as its law."[36] That is, through the visible flux of the morph's becoming, we discern a metaphysical and metastatically fixed and absolute essence of transformation and flux that undoes all our conceptions of human form and bodily animation. Here, what we might call the "law" of the morph—its dual status as both concrete and metaphysical, seductive in its continual movement and threatening in its absolute statis—is well summed up by T. S. Eliot, not only in the lines that begin this essay, but also in those that follow:

> I can only say, *there* we have been: but I cannot say where.
> And I cannot say, how long, for that is to place it in time.
> The inner freedom from the practical desire,
> The release from action and suffering, release from the inner
> And the outer compulsion, yet surrounded
> By a grace of sense, a white light still and moving,
> *Erhebung* without motion, concentration
> Without elimination, both a new world
> And the old made explicit, understood
> In the completion of its partial ecstasy,
> The resolution of its partial horror.[37]

In practice, of course, the morph has been put in the existential—if banal—service of car commercials and Hollywood narratives. In practice, it also hardly provokes on any regular basis the reflective conundrums of transformation and similitude posed by Magritte's paintings, or the self-reflexive meditations on human temporality and history of Reeves's extraordinary video work, or the questions posed by Bakhtin about the chronotopic entwining and mutual transformations of human being and the world. In practice, the morph's fantastic, uncanny achievement of reversible and nonhierarchical relations of similitude has little to do with democracy or representing real social problems that emerge from a given culture's racial, sexual, and specie-ial discriminations and much to do with dramatizing the existentially untenable myth of a heterogeneity that can be homogenized easily, without labor, without struggle, without violence, without pain. Furthermore, in practice, while the morph as a distinct digital figure still stands historically in the foreground of our vision, it is quickly being superseded by its functions in the background of our visible fictions. It is becoming normalized and thus increasingly invisible—just as the cinematic cut did before it.

Certainly, then, at this historical moment when we still can see the morph, we should interrogate its very existence and think through its meaningful relation to ours. And that relationship is one of both meanings and practices—for the digital morph changes us as much as we change it. That is, we have come to transform ourselves in its image of transformation and imagine our existence within its temporal imagination. Thus there is a correlation between the image of the morph and our morphological imagination.[38] It is hardly accidental that the historical moment in which "morphing" emerges as a foregrounded form of representing the human body as infinitely malleable is the same moment in which our own bodies have become subject to the technological transformations and (mythic) "quick fixes" found in cosmetic surgery and in the body-sculpting machines at the local gym. Both a cause and effect of the technological innovation that has given us the "morph" in all its temporal superficiality, we now foreground human existence as a material surface amenable to endless manipulation and total visibility. Faced with the visible representation and operations of an impossible meta-physical object, we fool ourselves into thinking we can conquer both our human flesh and temporality to get out of this world alive.

I have attempted here both a phenomenological and historical description of the "essential" structure and meaning of the digital morph as its novel objectification visibly emerges, in Eliot's words, as "both a new world and the old made explicit." Which is to say that a phenomenological description will always also be "qualified" by history and culture—in this case the Western history and culture of the writer (if not necessarily the reader). Yet oscillations and qualifications of meaning make the description no less accurate—just more partial. Unlike the morph itself, I exist as a being transformed not only *in* but also *by* time. And so the end of my own thought here turns back to its beginning—but with the difference of a reflexive comprehension that is historical. Eliot ends the poetic passage that I have quoted throughout this essay (and that speaks to the impossible and in-forming structure of movement and change in time that is "the still point of the turning world") by recalling history and celebrating the human mortality that makes of time a value. Against the "still point" as both absolute beginning and end to human animation and history, against its "completion" of the world's "partial ecstasy" and against its "resolution of its partial horror," he tells us:

> Yet the enchainment of past and future
> Woven in the weakness of the changing body,
> Protects mankind from heaven and damnation
> Which flesh cannot endure.[39]

To borrow from Eliot: whatever present fascination we may have with the metaphysical object that is the digital morph, the enchainment of past and future woven in the weakness of our changing bodies in-forms us with temporality and makes of us historical subjects. That is, our changing lives in time protect us (whether we will it or not) from the "still point" that attracts us as both origin and end, from the "heaven and damnation" of either absolute metastatic flux or absolute metaphysical fixity—neither of which, unlike our imagination, our flesh could possibly endure.

Notes

A much abbreviated version of this essay appeared as "Meta-Morphing," *Art+Text* 58 (August–October 1997): 43–45.

1. In the discussion that follows, emphasis is on digital morphing as it is used and appears within the context of photo-realism and is figured primarily around animated (or lived) bodies. Much of what I claim here about the uncanny qualities of the morph and its relation to human temporality and gravity cannot be mapped onto the metamorphoses of the "animated" film—whether digitally rendered or hand drawn. In the cartoon world, metamorphoses are the norm. In the discussion that follows, emphasis is on digital morphing as it is used and appears within the context of photorealism and is figured primarily around animated (or lived) bodies. For an excellent and detailed discussion on the impact of the digital on photo-realist cinema and computergraphic construction of "perceptual realism," see Stephen Prince, "True Lies: Perceptual Realism, Digital Images, and Film Theory," *Film Quarterly* 49, no. 3 (spring 1996): 27–37.

2. Reference here is to the "chronotope" (time-space) as conceptualized and applied to literature in the work of Mikhail Bakhtin, particularly his "Forms of Time and Chronotope in the Novel," in *The Dialogic Imagination: Four Essays by M. M. Bakhtin,* trans. Caryl Emerson and Michael Holquist (Austin: University of Texas Press, 1981), 84–258. As glossed by Holquist, the chronotope is an "utterly interdependent" ratio of time and space that provides "a unit of analysis for studying texts" as well as "an optic for reading texts as x-rays of the forces at work in the culture system from which they spring" (425–26).

3. For reasons of length, I am excluding from my discussion here elaboration

of the use of associative montage to create the continuities of simile and metaphor. One could argue such usage as aiming to constitute meaning as atemporal and conceptual. Nonetheless, just to further complicate the issue, one could also argue that such montage cannot escape time so easily: simile and metaphor always have a primary and secondary term and suggest a hierarchical relation between the two. Hence, in natural language or in a montage sequence, "My love is like a red, red rose" cannot be temporally reversed without a change in hierarchy and hence meaning. Here, albeit emphasizing there is no such thing as a film grammar, Jean Mitry, in *The Aesthetics and Psychology of the Cinema,* trans. Christopher King (Bloomington: Indiana University Press, 1997), is apposite: "Montage (which is the basis for film language) is nothing more than another form of language morphology where subject, verb, and complement have meaning only by virtue of their interrelationship" (127).

4. In early cinema, this visible yet transparent temporal/spatial gap emerges phenomenologically as the *surprising* and *abrupt* transformations of trick films wrought by "stop motion substitution": interruption of the "single" shot in camera to replace elements in the profilmic scene before resumption of shooting from precisely the same cinematic point of view; this "interruption" of the single shot could certainly be described as a "cut" in time, preservation of the continuity in spatial point of view providing the ground for the surprise of the temporal ellipses. Furthermore, recent research indicates "previously unperceived" actual out-of-camera editing (a "splice of substitution") in many of Georges Méliès's trick films. Thus Tom Gunning, in "'Primitive' Cinema: A Frame-Up? Or The Trick's on Us," in *Early Cinema: Space, Frame, Narrative,* ed. Thomas Elsaesser (London: BFI Publishing, 1990), asks: "Is Méliès not only a master of collage, but in fact the father of montage?" (98). At a somewhat later point in the history of cinema, this visible temporal gap is precisely what the normative practices of "classical" Hollywood editing attempted to elide through directing spectatorial attention to the narrative; hence, such editorial practices as cutting on an action.

5. Review of *The Wolf Man, CineBooks Motion Picture Review Guide,* on *Microsoft Cinemania* 1996.

6. See André Bazin, "The Virtues and Limitations of Montage," in *What Is Cinema?* trans. Hugh Gray (Berkeley: University of California Press, 1967), 41–52.

7. E-mail correspondence with the filmmaker indicates the film was loaded in the camera backward with the motor in reverse. Particularly telling here about the uncanny effect of reverse motion is Lawder's commentary: "I had shot the scene three times, all in regular forward motion, and still the cropping and camera angle was not right and I knew after viewing the third attempt that I had to shoot the scene again. Still I thought, this last shoot had some interesting material which I

wanted a quick second look at . . . so, rather than re-winding and screening the reel again, I thought let's simply flip the reverse switch on the projector . . . which is what I did, and to my amazement a totally unexpected set of sensations unfolded on the screen. *Necrology* was really born in that casual, instantaneous flip-of-the-switch decision and I returned to Grand Central Station with the Arriflex loaded backwards" (23 March 1998).

8. Reference here to "being" and "Being" derives from Martin Heidegger. For a gloss on the distinctions made, see *The Cambridge Dictionary of Philosophy,* ed. Robert Audi (London: Cambridge University Press, 1995), 317–19.

9. Of particular relevance here for its focus on the temporality and "chronotope" of the bildungsroman as well as its differences from previous forms, see M. M. Bakhtin, "The *Bildungsroman* and Its Significance in the History of Realism (Toward a Historical Typology of the Novel)," in *Speech Genres and Other Late Essays,* trans. Vern W. McGee, ed. Caryl Emerson and Michael Holquist (Austin: University of Texas Press, 1986), 10–59.

10. Indeed, it is telling that *Mortal Kombat* (1995), a film based on an eponymous computergraphic game filled with endless battles between "mighty morphin'" antagonists, deflects our attention from the morphs' impossible "mortality" in its title to the "playful" misspelling of "kombat."

11. This paradox and its relation to current figurations of ethnicity as a consensual form of identity and a commodity is further elaborated in Vivian Sobchack, "Postmodern Modes of Ethnicity," in *Unspeakable Images: Ethnicity and the American Cinema,* ed. Lester D. Friedman (Champaign-Urbana: University of Illinois Press, 1991), 329–52.

12. Jean Baudrillard, "The Hell of the Same," in *The Transparency of Evil: Essays on Extreme Phenomena,* trans. James Benedict (London: Verso, 1993), 122.

13. Michel Foucault, *This Is Not a Pipe,* trans. and ed. James Harkness (Berkeley: University of California Press, 1982), 32.

14. Ibid., 44.

15. I am indebted to Ilsa J. Bick for this insight about the palindromic nature of Odo's name and have extrapolated from this specific to a more general argument about the morph as a digital and cultural form.

16. Attributed primarily to the Stoics and Nietzsche, the doctrine of "eternal return" posits that "the same events . . . have occurred infinitely many times in the past and will occur infinitely many times in the future" and "stresses the inexorability and necessary interconnectedness of all things and events." As such, it is "antithetical to philosophical and religious viewpoints that claim that the world order is unique, contingent in part, and directed toward some goal." See *The Cambridge Dictionary of Philosophy,* 243.

17. This "reduction" of difference to what emerges as the "oneness" of movement can best be demonstrated by working in "correspondence mode" in a morphing application such as Elastic Reality; here one is able not only to see a fairly small number of "selected areas" of similarity between two images (i.e., eye, nose, mouth areas outlined) but also to see represented the vector lines "joining" them into a unitary area that will be "rendered" in the morph's movement.

18. Baudrillard, "The Hell of the Same," 122.

19. I am deeply indebted to Michael Renov for bringing this video to my attention.

20. T. S. Eliot, "Burnt Norton," in *Four Quartets* (New York: Harcourt Brace, 1943), 15. Much of this extraordinary poem might well serve as a gloss not only on human temporality and mortality but also on the morph's *unheimlich* revelations about the difference between Being and being. (For the epigraph that begins this essay, see pp. 15–16.)

21. Steve Seid, "*Obsessive Becoming:* The Video Poetics of Daniel Reeves," in *Works by Daniel Reeves,* http://www.mediopolis.de/transmedia/english, 13 May 1999.

22. "The Begotten," *Star Trek: Deep Space Nine,* production 510, written by René Echevarria, directed by Jesús Salvador Treviño (original air week: 27 January 1997).

23. There are many more complications to the plot than I can elaborate here. Most relevant, however, is that in previous episodes, Odo has been deprived of his shape-shifting abilities by the Founders of his home planet and regains them from the infant changeling as it dies in this episode's end; and that for various reasons, Major Kira has been a surrogate host for the embryo of the two human characters who are its biological parents.

24. Sigmund Freud, *Beyond the Pleasure Principle,* trans. and ed. James Strachey (New York: W. W. Norton, 1975), 43.

25. Bakhtin, "Forms of Time and Chronotope in the Novel," 113.

26. Ibid., 114.

27. Ibid.

28. Ibid.

29. Ibid., 115.

30. Ibid., 119.

31. Bakhtin, "The *Bildungsroman* and Its Significance in the History of Realism," 10, 19.

32. Ibid., 17–18.

33. Ibid., 23.

34. Ibid., 23–24.

35. Jorge Luis Borges, "Pierre Menard, Author of *Don Quixote*," trans. Anthony Bonner, in *Ficciones* (New York: Grove Press, 1962), 45–55.

36. Gordon Teskey, "Mutability, Genealogy, and the Authority of Forms," *Representations* 41 (winter 1993): 105. Teskey's argument is focused around Edmund Spenser's *Two Cantos of Mutabilitie* and its differences from the poet's *Faerie Queene.*

37. Eliot, "Burnt Norton," 16.

38. What is suggested here is a phenomenological history of the entailments of technology, representation, and the subject in constructing the temporal and spatial meanings of a specific life-world. See, as exemplary, Stephen Kern, *The Culture of Time and Space: 1880–1918* (Cambridge: Harvard University Press, 1983). Within a smaller compass specifically related to the historical relationship between morphing and other special effects technologies to the sense we have and make of our own bodies, see my own "Scary Women: Cinema, Surgery, and Special Effects," in *Figuring Age: Women, Bodies, Generations,* ed. Kathleen Woodward (Bloomington: Indiana University Press, 1999), 200–211.

39. Eliot, "Burnt Norton," 16.

8

Narratives of the Posthuman
After Arnold Cinema

ROGER WARREN BEEBE

The term "posthuman" has of late appeared with increasing frequency in critical writings.[1] It generally appears to signify some still rather imprecise melange of the postmodern (primarily as anatomized by Fredric Jameson), of the cyborg as elaborated by Donna Haraway, of Deleuze's (and Guattari's) postsubjective ethics, and, to a slightly lesser degree, of the critiques of humanism by Michel Foucault and Louis Althusser.[2] The common thread in these often divergent elaborations of the concept of the posthuman concerns the transformation of the subject through the emergence of new technologically capacitated forms of being and through the shift from productive to reproductive technologies. In a simultaneous movement in film studies (which has only occasionally been grounded in an awareness of these other contemporaneous cultural shifts), many scholars have posited a recent transformation of cinema, a transformation that is generally taken to concern spectatorship and narrative pleasure. Miriam Hansen, in her essay "Transformations of the Public Sphere," points to this radical change:

> The very category of *the* spectator developed by psychoanalytic-semiotic film theory seems to have become obsolete—not only because new scholarship has displaced it with historically and culturally more specific models but because the mode of reception this spectator was supposed to epitomize is itself becoming a matter of the past.[3]

Figure 8.1 On the set of *Terminator 2:* after encountering the T-1000, action hero Arnold Schwarzenegger is "floored." Photograph by Louise Krasniewicz.

Hansen identifies two primary and interrelated transformations—one in the academic *discourse* of film studies and one in the cinematic *apparatus* or *dispositif*.[4] This essay, however, proposes to explore a third coincident change within the cinema: namely, a transformation of the *structure* of Hollywood narrative film between the era of the 1980s action hero and the emergent Hollywood form of the 1990s, which I will term "the post-human cinema."[5] My subtitle mobilizes "narrative" in two distinct ways: it refers both to the narratives that we tell about an emergent posthuman cultural formation (with its posthuman subject and posthuman apparatus) and to the new cinematic narrative forms that are produced under and bespeak this new cultural formation.

Following the general theoretical trend in granting a certain centrality to the place of the technological in the emergent posthuman form, here I will examine a technological innovation—namely, the computergraphic morph (with specific reference to its use in James Cameron's *Terminator 2: Judgment Day* [1991])—as a figure for, and symptom of, the shift from the classic narrative model of the action hero film to a posthuman model. In its appearance as a visible rupture, the morph will serve as a hyperbolic dramatization or staging of the shift between the action hero film and the posthuman cinematic form.[6] Adapting and historicizing Roland Barthes's notion of the photographic "punctum" (a Latin term meaning "point," which Barthes uses to describe a localized rupture in the photographic image), I will examine the morph as a rupture in narrative (a sort of punctuation or suspension of narrative in a moment of sublime spectacularity) and as a rupture in a greater historical narrative (a way of marking a turning point on the way to a posthuman cinematic form and affective regime).[7] In my elaboration of the first of these two moments, I will explore the virtues and shortcomings of considering the morph as an instantiation of the "cinema of attractions" elaborated by Tom Gunning.[8] Gunning's influential concept, which is one of the touchstones of the recent shifts in film studies signaled by Miriam Hansen, has gained currency as *the* model for understanding such moments of nonnarrative spectacle (as a supplement to psychoanalytic models of narrative and spectatorship) and thus begs comparison to the discussion of the morph as punctum developed in this essay.

Although the morph is the central focus of the essay (and the central pivot in the historical narrative that it will develop), it is itself only the unstable moment of rupture or a way of dramatizing that moment where a new form begins to emerge and not, finally, the stabilized form of posthuman narrative. The morph merely functions, through its disruptive pleasures, to unsettle the terrain of Hollywood narrative allowing for the emergence of a new form. Therefore, I will end this essay with a brief exploration of the emergent posthuman cinematic form as manifested in such films as Steven Spielberg's *Jurassic Park* (1993), Roland Emmerich's *Independence Day* (1996), and *Godzilla* (1998) as the aftermath of the disruption of the morph in *T2.*

The Punctum of the Morph

In a prescriptive addendum to Miriam Hansen's descriptive analysis of the present state of film studies, Linda Williams has recently suggested:

> While the psychoanalytically derived models of vision that have dominated film theory in recent years have enabled the analysis of certain kinds of *power*—the voyeuristic, phallic power attributed to a "male gaze"—they have sometimes crippled the understanding of diverse visual *pleasures* despite the importance and prevalence of the very term *pleasure* in them.[9]

In fact, posthuman narrative seems to demand that we begin to explore these "diverse visual pleasures" as new kinds of (disruptive) pleasure emerge surrounding the figure of the morph in *Terminator 2.*

Although I think it is important to recognize, as have most other commentators, that in *T2* there is a sustained effort to make Schwarzenegger's older-model Terminator a now kinder, gentler cyborg more suitable for identification than his/its first incarnation in *The Terminator* (1984),[10] this gesture seems to be merely compensatory for what would otherwise be a noticeable void in a Hollywood action film: that is, the lack of a likable (or at least admirable) adult (male) (human) figure—like Kyle Reese in the first Terminator film—who can serve as the spectator's proxy for the experience of the film.[11] Sarah Connor (Linda Hamilton), with her new hard-bodied look, might have helped fill that void were she not doubly pathologized in *T2.* First, she is pathologized by the medical establishment, locked up in an insane asylum (wrongly) for her presumed paranoid hallucinations of the future and kept locked up (more justifiably) because of the violence which her mania elicits.[12] Second, and more crucially, she is pathologized by her son John, the future savior of humanity, who opposes her desire to kill those responsible for the computers that will eventually attempt to destroy the human race (as well as anyone else who might happen to interfere). When Sarah fails in her attempt to kill Miles Dyson, the scientist who invents the Cyberdyne system that will destroy the world, she heatedly delivers a long tirade against the destructive nature of Man/men to him: "You think you're so creative. You don't know what it's like to really create something. To create a life. To feel it growing inside you. All you know is how to create death and destruction and. . . ." John is noticeably embarrassed and is shown in a reaction shot hiding his face in his hands as she says, "To create a life." He finally interrupts her, offhandedly dismissing this tirade as a product of the fervent excesses of her (feminine, maternal) outrage: "Mom, we need to be a little more constructive here, okay?" These constant gestures of pathologization in the film undermine Sarah's credibility as a center of identification.

Ultimately, John Connor may be the closest thing we get to a *human* proxy in this film (especially for the teenage boys who make up a large sector of the audience for action hero films). Certainly one thread of the narrative is about John's slow coming into manhood, but (despite being the future savior of humanity) his heroism in the film is limited to displaying his savvy at manipulating a handheld computer to rip off an ATM and then to enter Cyberdyne's inner sanctum. At the end of the film, he remains a teary-eyed teen who refuses to recognize the necessary sacrifice of his buddy/dad/protector, the T-101 (Schwarzenegger).

Although this lack of a clear human center for identification might have proved to be the film's downfall, *T2* is largely able to eschew this potential problem through the appearance of the distinct and remarkable spectatorial pleasures that emerge around the metal-morphing T-1000 (and morphing technology in general—particularly because the film was released at the moment when it was the latest technological marvel). These are pleasures characterized by their disruptive nonnarrativity. Thus it seems useful to begin to explore the effects of the morph by exploring Barthes's notion of the punctum. Unfortunately, as Jameson suggests, "Barthes' analytic concept is a necessary starting point, but only that."[13] But, as "a necessary starting point," his elaboration of the punctum provides the ground for an understanding of the disruptive pleasures of the morph.

The punctum, as developed in Barthes's *Camera Lucida,* describes a complex of meaning ranging from the affective notions of the "wound . . . the mark . . . the prick" to the concepts of "punctuation" and rupture.[14] In the split in this definition (between the affective pole and the rhetorical pole), it is clear that the punctum operates on two levels. First, the punctum creates an intense affect in the spectator that Barthes describes as "an internal agitation, an excitement . . . the pressure of the unspeakable which wants to be spoken."[15] It is this intensity of affect that initially provoked Barthes's inquiry and proves to be the central focus of *Camera Lucida.* Second, this intense affect challenges the unity of the photograph: "A detail overwhelms the entirety of my reading."[16] Here Barthes echoes the formulation of the sublime attributed to Longinus. While the well-crafted work for Longinus is "the hard-won result not of one thing or of two, but of the whole texture of the composition," other works offer the disruptive power of the sublime, which "flashing forth at the right moment scatters everything before it like a thunderbolt."[17] Indeed, Barthes even echoes Longinus's metaphorical language of the thunderbolt in calling the punctum "lighting-like"[18] and "a fulguration."[19] Although it is initially useful to isolate the affective and rhetorical levels on which the punctum functions in still photography, it is ultimately necessary to reunite them to begin to see the importance of the punctum in understanding the challenge that the morph represents for cinema: the punctum presents an affective challenge of a eruption of sublimity that upsets the normal rhetorical functioning of the photographic signifier.

Although Barthes limits his exploration of the punctum to the specificities of the still photograph, a cinematic analogy is apt, but it requires some translation. Whereas Barthes reads the punctum's rhetorico-affective challenge as a spatial disruption (primarily to the unity of composition, although also to the unity of interpretation), the cinematic punctum functions additionally (and perhaps primarily) as a temporal disruption, an interruption of the flow of narrative and a suspension of linear time. In its volatile, affective disruptions, the cinematic punctum offers a type of radically non- or antinarrative pleasure, which provides the ground for an analysis of the morph.

These initial descriptive gestures are given substance through a coupling with Jameson's exploration of the reinvigoration of the affects of the sublime under postmodernism. His inquiry covers much of the same affective territory as does Barthes's elaboration of the punctum, but within a much more precisely historicized context that is the necessary supplement to Barthes's musings. Whereas Barthes never attempts to ground his investigation in history or in the construction of the subject of a historical moment (hence, perhaps, Jameson's critique), Jameson makes explicit the link between the postmodern upheaval of subjectivity, the shift from productive to reproductive technologies under late capitalism, and this new affective regime.[20] This regime, which emerges with the (so-called) postmodern "waning of affect," accompanies a contemporary reinvigoration and transformation of the affects of the sublime (affects, as we have already seen, which are strongly linked to Barthes's punctum).[21] In Jameson's terminology (adapted partially from Lyotard), the affects of the postmodern can best be described as "joyous intensities" or a "free-floating euphoria" that emerge after the crisis of the bounded subject. Jameson's passages on the contemporary transformation of affect are explored at length in Vivian Sobchack's remarkable final chapter of *Screening Space,* specifically with reference to the use of the special effect in science fiction films.[22] As she writes in the section "The Transformation of Special 'Affect' into Special 'Effect'" (a title that almost says it all), "although special effects have always been a central feature of the SF film, they now carry a particularly new affective charge and value."[23] The irruptive pleasures of the morph are one historically specific form that this "new affective charge and value" takes. Sobchack echoes Jameson in her elaboration of the specific character of this affect, suggesting that "the primary sign-function of SF special effects seems to be precisely to connote 'joyful intensities,' 'euphoria,' and the 'sublime.'"[24] Thus understood as a combination of

Jameson (as interpreted by Sobchack) and of Barthes, the cinematic punctum implies a momentary explosion of sublime affect produced by the punctuating presence of the morph and an associated moment of both narrative and epistemic disruption that emerge out of a very concrete historical and cultural context. The morph as a specific realization of the cinematic punctum thus renders visible the reshaping of the subject and of the pleasures produced by an older cinematic/narrative and cultural order as its disruptive pleasures in *T2* unsettle the (narrative) terrain—thus allowing for the emergence of the posthuman cinema.

Punctum in Act(ion)

Although the commercials for *Terminator 2* displayed some of the low-tech violence (explosions, shooting cops in the kneecaps) and the one-liners that had become requisite fare in the action hero film, what was truly foregrounded in the massive paratext for *T2* was the high-tech wizardry of the T-1000.[25] Whereas the first Terminator film relied on explicit and repeated violence and the belabored action of Schwarzenegger's hulking form much more than on special effects (which were limited to more traditional makeup and robotics work), *T2* was explicitly (if not exclusively) a showcase for morphing (and, by extension, for the technologies that produced it). The T-1000 is not the Herculean (or Schwarzeneggerian) hulk of its predecessor, where its mere quantitative difference in size would indicate its qualitative difference and superiority. On the contrary, the T-1000 in its human form is extremely slight of build. Yet in the disruptive force of the punctum it produces, it partakes of what Paul Virilio refers to as the "aesthetics of disappearance," where visibility and size are no longer the principal axes of the expression of distinction or power. As Virilio notes (in language that is interestingly appropriate for a discussion of the shift away from the action hero film), "At the close of our century . . . we live in the beginnings of a paradoxical *miniaturization of action*."[26] The fluid and effortless morphing of the T-1000 stands in stark contrast to the unwieldy and laborious spectacle of Schwarzenegger's action, which was requisite fare in the action hero film of the 1980s.[27]

Whereas human physical action constantly requires bigger, faster, and more, the morph runs counter to these trends by initiating not only the "miniaturization of action" described by Virilio but also a coincident temporal protraction. In its less theoretically interesting moments, this slowing down derives rather directly from typical horror film chases where the killer's slow and deliberate pursuit intensifies the terror. This structure is a major element in *The Terminator* and, as we shall see momentarily, still has some role in the sequel. However, in addition to the simple suspense created by the temporal implacability of the horror film's stalker, the presence of the morph in *T2* (and elsewhere) also results in a suspension: a suspension in the progress of the narrative, a suspension, as it were, of suspense itself (and its concern for futurity) as the sublime punctum of the morph momentarily distracts our attention from the action in progress.

One of the more notable examples in *T2* is the scene in the state hospital at Pescadero where Sarah Connor and her would-be emancipators, John Connor and Schwarzenegger's T-101, are pursued by the T-1000. The sequence operates through a classic multiple cross-cutting structure (among Sarah, the hospital employees, John and the T-101, and the T-1000) that frequently employs the first kind of temporal protraction referred to earlier. This form predominates until the three protagonists are united and the T-101 has defeated all of their human pursuers. As they regroup and prepare to exit the mental hospital, they see the T-1000 approach a locked gate that separates it/him from them. It/he stops momentarily before the gate, eliciting reaction shots from the three fugitives as well as from the staff psychiatrist. Having already experienced the punctum of the morph, we too now await its imminent (re)appearance, and our concern for the development of the chase is put temporarily on hold. The characters themselves seem to have completely forgotten their escape-in-progress at this point, and they sit still waiting for something to happen. This is a first suspension of the suspense of the chase as it is displaced by a different kind of suspense: the suspense of awaiting the appearance of the morph. The second, more radical, suspension takes place as the T-1000 slowly morphs around and through the bars, which seem simply to slide through it/him. This second suspension abandons all concern for futurity, is fully absorbed in the presence and in the becoming of the morph, and the narrative is fully ruptured by the sublime process of the morph. The (pro)tension built up during the morph is finally resolved by a gag, as a close-up shows the T-1000's pistol getting stuck between the bars through which it/he has just effortlessly passed. Only then does the chase recommence with a relatively standard action sequence (itself broken up by a number of smaller morph-puncta).

This type of temporal protraction and suspension of suspense is explicitly figured as one of the most remarkable pleasures of the film in its promotional paratext. In an advertisement encouraging viewers to purchase their own copies of the video cassette, the father of the representative household is depicted as using the capacities of his remote control to further this temporal protension by rewinding a morphing scene (specifically, the scene where the T-1000 enters a helicopter outside the Cyberdyne building by pouring itself through a hole in the windshield). He cheerily announces, "I've gotta see that again," before rewinding the scene not just once but twice. As if to further emphasize the specific mechanism for enjoying the morph, the sequence also includes a close-up of the LCD display on the VCR that reads "REW." The advertisement can thus be seen as a sort of "instructional video" teaching us how best to experience these remarkable new pleasures. Endorsing this kind of interruption of the flow of the narrative under the order of classical narrative is almost unthinkable, whereas here it is shown as being integral to the pleasure of the viewing experience. Thus the viewer is specifically encouraged to desire—and is shown as desiring—the kind of temporal protraction and suspension of suspense in the home viewing experience that I am positing as a key component of the distinct cinematic pleasures of the morph-punctum as narrative interruption, pleasures that present a radical (if only intermittent) challenge to the general force and pleasure of narrativity.

The Morph and the Attraction

This type of extranarrative eruption associated with the morph as cinematic punctum could certainly be related to one of the "historically and culturally more specific models" to which Miriam Hansen refers.[28] Most immediately relevant is Tom Gunning's (increasingly canonical) work on the "cinema of attractions." Gunning attempts to demonstrate the relation between cinema and earlier forms of nonnarrative spectacle that created "a cinema of instants, rather than developing situations," a phrasing that parallels the language of my discussion of the interruptions of the morph.[29] Gunning's theory represents a break from traditional conceptions that grounded cinematic pleasure in the seamless functioning of narrative illusion, offering instead a more heterogeneous notion of the sources and types of cinematic pleasure. He suggests the relevance of this historical insight for contemporary film theory by placing the "attraction" in a historical dialectic:

> Even with the introduction of editing and more complex narratives, the aesthetic of attraction can still be sensed in periodic doses of nonnarrative spectacle given to audiences (musicals and slapstick comedy provide clear examples). The cinema of attractions persists in later cinema, even if it rarely dominates the form of a feature film as a whole.[30]

Clearly, then, the suggestion could be made that the morph as "attraction" represents yet another of the "periodic doses of non-narrative spectacle." However, although I agree that the pleasures of *T2* I have been describing, like those of the cinema of attractions, rely on a series of extranarrative "shocks" or "thrills" related to a new technologically enabled form of perception, I would be cautious about the simple subsumption of the morph-punctum under the concept of the attraction. As liberating as the suggestion of another system of pleasure functioning alongside (and sometimes counter to) the traditionally conceived system of narrative pleasures and as productive as understanding the historical continuity of these non-narrative pleasures may be, such gestures may ignore the cultural specificity of the morph as a new kind of attraction developing out of a concrete historical formation or with a unique set of affective, ethical, and political problematics. It is only through an examination of this cultural specificity that we can begin to understand the significance or importance of the morph, as I began to suggest earlier in supplementing Barthes's notion of the punctum with the concrete historico-affective configuration of postmodernism. In this spirit, then, I will now (more than a half decade since the T-1000 morph erupted onto the scene in *Terminator 2*) explore the morph not as *the* new form of cinematic pleasure but rather as merely the literal passage point from a specific narrative form of the 1980s, the action hero film, to a posthuman narrative form in the cinema of the 1990s.[31]

After the "Last Action Hero"

The action hero film provides the ground from which the morph violently surfaces in *T2*. From around the time of the release of the first Terminator film until the release of *T2*, action hero films enjoyed an unchallenged position in the movie market.[32] The pumped-up images of Schwarzenegger and Stallone were presumably the ego ideals for a great number of American men and boys. Some might contend that they still are. However, as the "joyous intensities" of the cinematic punctum intermittently shift interest away from the plights of these (super)humans in *T2*, a reconfiguration of the narrative field is set in motion.

Although one of the initial promises of *T2* was that a new, pumped-up Linda Hamilton might prove to be a transformative figure in challenging the male stronghold of the action hero film, what actually happened (as attested to in part by Hamilton's return to relative obscurity)[33] is that the genre itself was rendered no longer viable.[34] The old T-101's clumsy cyborg mechanics (i.e., the human body, even an "ideal" one like Schwarzenegger's) are displaced (and, as we shall see momentarily, not simply replaced) by the blindly embodied intentionality of the metal-morphing T-1000. This shift occurs simultaneously along two related fronts: first, a move away from the star system of the 1980s (i.e., the decreasing popularity of the action hero vehicle), and second, a transformation of the technological modes of cinematic production (i.e., the morph as the herald for new uses of digitalization in the film image). Beginning to elaborate the first of these should give some purchase on an understanding of the second.

Given the centrality of the metal-morphing T-1000 in *T2,* it might seem surprising that the actor commonly credited with playing the T-1000 (Robert Patrick) quickly returned to relative obscurity.[35] However, this disappearance is actually a logical result of the narrative and affective challenge of the film. It is not the humanly embodied Patrick but rather the T-1000 (i.e., the morph and morphing technology) that is the center of interest in *T2.* The film (and the requirements of its specific narrative form) are indifferent both to Patrick's material existence as a historical individual or human subject and to his existence as an actor or a potential star and action hero. Just as the T-1000 has no body, it is also nobody, a constant shape-shifter that resists easy classification, an embodied agency that frustrates the Cartesian separation of mind and body. In fact, I have avoided until now referring to Patrick at all because the T-1000 is actually embodied by several different actors in the course of the film. In addition to Patrick, it is given human shape by the actress playing John's foster mother, by the twin of the actor playing the guard at the insane asylum, and, finally—and quite provocatively—by Linda Hamilton's real-life twin. Although one may find Patrick's performance memorable or interesting (as does Albert Liu in his discussion of Patrick's preparatory study of insect locomotion in his remarkable essay "Theses on the Metal Morph"),[36] Patrick succeeds most when he disappears: that is, when his historical and personal reality is effaced as he "becomes," through the aid of the morph, the technological attraction that we have come to see.

Unlike Patrick's nonstar status, Arnold Schwarzenegger's star image still looms large in the film, even if he is not set up as a human protagonist suitable for identification. Although his character's inability to grasp certain basic modes of human communication makes him the butt of frequent jokes, his years of box office dominance still provide him with a certain aura. The film tries intermittently to tap into this star quality and to resist the radical shift instigated by the disruptive pleasures of the morph by evoking the standard pleasures of the action hero genre through Schwarzenegger's various low-tech heroics and through his wooden and self-conscious one-liners (which, unlike the deadpan and almost campy witticisms of the less "user-friendly" Terminator of the first film, frequently fall flat). These attempts, however, and the pleasures that they offer (if one does indeed find them pleasurable) are so frequently ruptured and interrupted by the puncta of the morphing scenes that the damage is already done: after Schwarzenegger's battle with the morph, the action hero film cannot continue as before. And in the allegory of the film, this damage is figured explicitly in the tattered body and bared (and now somewhat pitiful or pitiable) mechanics of the T-101.

In this light, we can understand why *Last Action Hero* was such a disastrous flop: it is not because its diegetic diagnosis of the crisis in Hollywood cinematic form (occasioned by the computergraphic challenge to the action hero) was incorrect. Rather, its diagnosis came too late, simply providing a postmortem of the action hero who had effectively been "terminated" in *T2.* By becoming the effective and affective center of interest and pleasure, the T-1000 is the real victor in the showdown between it and the T-101, despite the narrative outcome that attempts (unsuccessfully) to return us to both a comforting humanism and a tempered technophobia.

Posthuman Cinema: The Aftermath of the Morph

After this violent upheaval occasioned by the eruption of the morph, a new stable narrative form slowly emerges. Whereas in *T2* the pleasures of the T-1000 derive from the affects of the punctum, after *T2* this cinematic punctum (or at least its realization in the form of the morph) becomes itself displaced. That is, the morph is subsumed and banalized as part of the greater digital domain.[37] In simple terms, morphing as *the* new technology—or at least as the new extension of a technology that had been used in feature films at least since *Tron* (1982) and *The Last Starfighter* (1984) and in this form since Cameron's *The Abyss* (1989)—was clearly destined to become just another moment in the historical litany of technological innovations (color, sound, Cinerama, CinemaScope, 3-D, Smell-O-Vision, etc.). But beyond the process of banalization brought about through the morph's overexposure in various media (features, music videos, commercials), even the morph's use as the synecdochic stand-in for the more global processes of film digitization seems increasingly misconceived, since the morph represents only a momentary point of disruption and not a new form of effects-based narrative. A great number of the films with the highest domestic box office grosses in the years since *T2*—notably *Jurassic Park, Forrest Gump* (1994), *Toy Story* (1995), and *Independence Day*—begin to suggest a new model of narrative with a radically transformed role for digital effects. Although these films base a large part of their draw on the fact that they use computergraphic technologies, these technologies serve more as the *ground* for a new narrative form than as a means of upping the ante within the narrative codes of the older narrative form of the action hero film. This is where we begin to see the newly stable narrative forms of the *posthuman cinema. Jurassic Park* (or, more properly, nearly a billion dollars in international box office receipts) shocked Hollywood into the realization that a blockbuster need not be either a star vehicle or even a narrative centered on the plights and adventures of a single human subject.[38] The lack of a strong central (human) star in *Jurassic Park* and in those films of which it is the prototype results in a dispersal of the narrative and a multiplication of narrative centers. In this emergent posthuman cinema, Laura Dern, Sam Neill, Jeff Goldblum, Richard Attenborough, Samuel L. Jackson, Wayne Knight, and a couple of cute kids serve as the lesser human"stars" of one of the highest-grossing films of all time.

Jurassic Park bears comparison to *Star Wars* (1977), which seems at first glance to be its obvious precedent as an overwhelmingly successful, effects-heavy, star-light film. But unlike the cast of nonstars in *Star Wars,* none of the cast members of *Jurassic Park* are even potential stars—or if they are, the film's success is indifferent to that potential just as *T2* is indifferent to Robert Patrick's possible star status. Although *Star Wars* made a superstar of Harrison Ford and a household name of Mark Hamill and Carrie Fisher, the casting of *Jurassic Park* appears to follow a completely different model in which the actors are already established character actors whose functioning as stereotyped bit players (the grandiose old dreamer, the stuttering oddball mathematician, the middle-aged sourpuss–father manqué, etc.) almost precludes their future emergence as stars. Additionally, the central human "star" of *Jurassic Park* (if it is even possible to isolate one) is not some young, rising star but Sam Neill, a middle-aged journeyman actor whose fortunes seem unrelated to the success of the film. And again unlike *Star Wars,* furthermore, the sequel *The Lost World: Jurassic Park* (1997) jettisons most of the characters from the original, featuring only Jeff Goldblum's quirky chaos theorist (with a brief appearance by Richard Attenborough). We return to the *Star Wars* films to see the continuing (and soon, the previous) exploits of the characters we have come to care so much about, but when we return to *Jurassic Park,* we simply want to see more of the computergraphic dinosaurs. With the passage from a character-centered model for the hi-tech blockbuster to an effects-centered model, we pass from a "human" cinema to a posthuman one.

In films that follow this posthuman narrative model derived from, or exemplified by, *Jurassic Park,* no longer is it the punctuating morph that is primarily actualized by computergraphic technologies, but rather a whole new form of narrative enabled by them. While this form is clearly still emergent (as attested to by the summer 1998 release of hoped-for blockbusters *Deep Impact, Godzilla,* and *Armageddon* as modeled on slight permutations of this same formula), some new configurations are becoming increasingly clear.[39] Most visibly, there seems to be a proliferation of teamwork narratives that attempt to fill the central void created by the death of the action hero: for example, in relation to films released in 1995, *Toy Story, Crimson Tide, Congo, Apollo 13, Mighty Morphin' Power Rangers, Species, Babe, Mortal Kombat, Jumanji,* and *Hackers.*[40] This list should begin to make clear that while the posthuman cinema certainly includes its fair share of SF (many featuring the canonical figure of the cyborg), the posthuman cinema is about more than just these narrow generic issues: this emergent posthuman form can also accommodate animated toys, barking stiff-lipped military men, and cute talking pigs. It should also be clear from the inclusion of films such as *Apollo 13* and *Crimson Tide* in this list (both of which use CGI effects primarily for the simple task of creating exterior shots of their respective rockets and submarines) that the computergraphic novum is not simply determinate of the posthuman cinema, even if it provided the impetus for its emergence. Additionally, and perhaps surprisingly, in many of these films there has been a persistence of (or even a recrudescence of) the technophobic narrative of the uneasy sort found in *T2* where high-tech effects punctuate (and complicate) essentially luddite narratives. The new technophobic teamwork narrative found in most of the films—again following the model of *Jurassic Park*—is, in brief, the story of a group of humans who must unite to destroy some evil technology or its progeny technological breakdowns in *Apollo 13* and *Crimson Tide,* digitized and/or morphing bad guys (and gals) in *Mortal Kombat, Species,* and *Jumanji,* techno-capitalistic hubris in *Congo* and *Species* and its youthful parodic embodiment in *Jumanji,* and so on. It seems that this persistent technophobia might indicate that we are still in the aftermath of *T2,* which, in true Hegelian fashion, contains the contradiction that was sublated in the emergence of this new form.[41]

The use of the term "posthuman" to describe this type of narrative is somewhat hyperbolic: there is no simple transcendence of our reliance on the human protagonist after the decline of the action hero. This rhetorical gesture, however, marks the passage of my theorization of the posthuman from the diagnostic to the prognostic as I turn to the future of this emergent form. If the action hero film has generally been understood as being allied to Reagan-era masculinity (as in Jeffords's *Hard Bodies*), then positing a shift away from this model can also be seen as a move in the essay from the descriptive to the prescriptive. One of the utopian possibilities offered by my construction of the posthuman cinema is that these more diffuse posthuman narrative forms may allow for the emergence of a notion of the *collective* (of which the "team" of humans or monsters might be the emergent form) that might supersede the role of the isolated individual human subject (the legacy of previous star systems and narrative structures from film noir to the action hero). At least two of the films in the foregoing list explicitly embrace this utopian potential: *Hackers,* which features the emblematic line of dialogue "I may not be good enough to beat you—but *we* are" in both its theatrical and television promos, and *Mighty Morphin' Power Rangers,* which proclaims both on its posters and on its video cassette box "The Power of Teamwork Overcomes All." Despite its utopian possibility (which also figures in much of the discourse on the posthuman), this emergent form remains fraught with an equal amount of dystopian potential. One of the more obvious of these dystopian aspects of the posthuman is that the shift away from the 1980s star system may simply represent Hollywood's attempt to overcome its reliance on the troublesome human labor of stars. (Indeed, much of the panic over the use of computergraphic technologies seems to center on this possibility of the succession of actors and acting.) Of course, this strategy has already begun to fail as, ironically, a number of "star animators" have emerged who require the same costly investment in human labor as the action heroes that they helped to displace. So although the posthuman cinema may have shifted away from a politically despised model (Reagan-era masculinity), it remains nonetheless a contested and troubled terrain. This exploration can hopefully serve as a cognitive map of that terrain, which perhaps will allow us to intervene in its continued evolution.

Postscript: The Prescient Morph

The ambivalent victory of the morph (as a precursor of these new technologies and new technologically enabled forms) over the older form of the action hero (and the action hero film) can actually be seen retrospectively prefigured in a certain unpacking of the final scene of *T2.* With the melting and commingling of the bodies of the T-101 and the T-1000 in the vat of molten iron, the whole complex of interpretations supported by the two Terminators is fused and confused: namely, film and digital technologies, technophobia and technophilia, the human and posthuman. The final morph is a morph to nothing, a swan song for this appearance of the cinematic punctum, as the T-1000 slowly becomes the blank white surface of the screen. The metal-morph's sacrifice ultimately "morphs" cinema itself. And Schwarzenegger's thumbs-up as he follows the morph into the molten alloy lets us know that even though this time he won't "be back," we'll learn to get on fine without him.

Notes

1. See Judith Halberstam and Ira Livingston, *Posthuman Bodies* (Bloomington and Indianapolis: Indiana University Press, 1995); Tiziana Terranova, "Posthuman Unbounded: Artificial Evolution and High-Tech Subcultures," in *Future Natural: Nature/Science/Culture,* ed. George Robertson (London and New York: Routledge, 1996); Claudia Springer, *Electronic Eros: Bodies and Desire in the Postindustrial Age* (Austin: University of Texas Press, 1996); and Tom Cohen, *Antimimesis from Plato to Hitchcock* (Cambridge: Cambridge University Press, 1994).

2. The key texts are Fredric Jameson, *Postmodernism, or The Cultural Logic of Late Capitalism* (Durham, N.C.: Duke University Press, 1991); Donna Haraway, "A Cyborg Manifesto: Science, Technology, and Socialist-Feminism in the Late Twentieth Century," in *Simians, Cyborgs, and Women: The Reinvention of Nature* (London and New York: Routledge, 1991); Gilles Deleuze and Félix Guattari, *Anti-Oedipus: Capitalism and Schizophrenia,* trans. Robert Hurley, Mark Seem, and Helen R. Lane (New York: Viking Press, 1977), and *A Thousand Plateaus* (Minneapolis: University of Minnesota Press, 1987); Michel Foucault, *The Order of Things: An Archaeology of the Human Sciences* (New York: Vintage Books, 1970); and Louis Althusser, *For Marx,* trans. Ben Brewster (London and New York: Verso, 1969), and *Lenin and Philosophy,* trans. Ben Brewster (New York: Monthly Review Press, 1971).

3. Miriam Hansen, "Transformations of the Public Sphere," in *Viewing Posi-*

tions: Ways of Seeing Film, ed. Linda Williams (New Brunswick, N.J.: Rutgers University Press, 1994), 135.

4. The term *dispositif,* for which "apparatus" is the standard but insufficient English translation, refers not simply to the technological apparatus of the cinema (in French, *appareil*) but rather to the entire arrangement of social, psychic, economic, ideological, and mechanical circumstances that come into play in the production, postproduction, distribution, projection, and consumption of a film. The transformation of the notion of the subject could therefore be considered as a part of the more general transformation of the apparatus.

5. The "action hero film" to which I will make reference for the rest of this essay should not be confused with the much more broadly defined "action" or "action-adventure" genre. By "action hero film" I wish to indicate a number of films from Reagan-era Hollywood best exemplified by a large majority of the Arnold Schwarzenegger vehicles after *The Terminator* and the Sylvester Stallone vehicles after *First Blood* (1982), and by *Die Hard* (1988) and its sequels. For a more thorough description of the boundaries of this subgenre, see Susan Jeffords, *Hard Bodies: Hollywood Masculinity in the Reagan Era* (New Brunswick, N.J.: Rutgers University Press, 1994).

6. Although the morph seems generally to represent a clean break or rupture (language that I will occasionally use to dramatize the change in narrative regime), more nuanced language is ultimately required that does not suggest such a clean supersession of the older narrative regime. Raymond Williams's vocabulary of "emergent," "residual," and "dominant" cultural forms seems apt. In this vocabulary, the essay proposes that the posthuman cinema is an emergent form (now dominant), whereas the action hero film is a residual form from an older era.

7. Roland Barthes, *Camera Lucida: Reflection on Photography,* trans. Richard Howard (New York: Hill and Wang, 1981).

8. See particularly Tom Gunning, "The Cinema of Attractions: Early Film, Its Spectator, and the Avant-Garde," in *Early Cinema: Space, Frame, Narrative,* ed. Thomas Elsaesser (London: BFI, 1990), 56–62.

9. Linda Williams, "Corporealized Observers: 'Visual Pornographies and the Carnal Density of Vision,'" in *Fugitive Images: From Photography to Video,* ed. Patrice Petro (Bloomington and Indianapolis: Indiana University Press, 1995), 6.

10. See Jeffords, *Hard Bodies,* or Tom Cohen, "Coda: Posthumanist Reading," in *Anti-mimesis from Plato to Hitchcock,* 260–64.

11. The first Schwarzenegger scene in *T2* plays with the fact that we actually identified with the Terminator in the first film. In this scene, we do not know yet (except through a familiarity with the film's paratext—commercials, reviews, etc.) that Schwarzenegger's Terminator is the "good guy" in this film: we are simply told

in the opening monologue that there are two Terminators—one good, one bad. Nonetheless there is a great amount of (admittedly complicated) pleasure to be had as we watch him destroy a group of bikers in order to get their clothes and motorcycle, pleasure that seems at least in part to issue from identification. In this light, the transformation of his Terminator from the first film to the second is substantially more (ideologically, structurally) complicated than it might first appear.

12. This second source of her institutional pathologization is still somewhat questionable given the great efforts that the film makes to demonstrate the incompetence or moral bankruptcy of the various hospital employees. This excuses her violence somewhat, but not fully, as the film continually explores the excesses of her rage.

13. Fredric Jameson, "On Magic Realism in Film," in *Signatures of the Visible* (New York and London: Routledge, 1992), 236.

14. Barthes, *Camera Lucida,* 26.

15. Ibid., 19.

16. Ibid., 49.

17. Longinus, "On the Sublime," in *Critical Theory since Plato,* ed. Hazard Adams (Fort Worth, Tex.: Harcourt Brace Jovanovich, 1992), 76.

18. Barthes, *Camera Lucida,* 45.

19. Ibid., 49.

20. Jameson, *Postmodernism,* 35–38.

21. There seems to be some confusion about the precise meaning of "affect" in contemporary theory. Jameson relates affect to a depth model of subjectivity that is outmoded with the emergence of postmodernism: hence the postmodern "waning of affect." Deleuze, on the other hand, considers affect as a simple state of vectors acting on bodies (almost like billiard balls), which is not contingent on the modern(ist) model of the subject. This latter kind of affect clearly persists in the pleasures of the sublime interruptions of the morph.

22. Vivian Sobchack, *Screening Space: The American Science Fiction Film* (New Brunswick, N.J.: Rutgers University Press, 1997).

23. Ibid., 282.

24. Ibid., 283.

25. The connection of various affects to violence seems to echo the traditional distinction made between the affects of horror (fear, terror) and those of science fiction (fascination). (For more on this differentiation between horror and SF see chapter 1 of Sobchack's *Screening Space.*) Interestingly, however, the morph-punctum of *T2,* in its relation to the sublime, incorporates an element of terror that is typically absent in accounts of classic SF. Not surprisingly, Barthes flatly

rejects the term "fascination" in the search to name the affect of the punctum early in *Camera Lucida* (18).

26. Paul Virilio, *Speed and Politics,* trans. Mark Polizzotti (New York: Semiotext[e], 1986), 140.

27. A productive theoretical analogy for this moment of rupture instigated by the morph (referred to in the title of this section) might be, using the Lacanian/Žižekian binary "act/action," to consider Schwarzenegger's "masculine" action as opposed to the "feminine" act of the morph, which presents a rupture and (momentary) transcendence or suspension of the symbolic. As Žižek lays out the problematic: "The act differs from the active intervention (action) in that it radically transforms its bearer. . . . after an act, I'm literally 'not the same as before.'" Slavoj Žižek, "Why Is *Woman* a Symptom of Man?" in *Enjoy Your Symptom! Jacques Lacan in Hollywood and Out* (New York and London: Routledge, 1992), 44. The parallel between this moment of rupture occasioned by the act and the shift in cinematic/symbolic form occasioned by the punctum of the morph merits further consideration.

28. Hansen, "Transformations of the Public Sphere," 135.

29. Tom Gunning, "An Aesthetics of Astonishment: Early Film and the (In)Credulous Spectator," in *Viewing Positions: Ways of Seeing Film,* ed. Linda Williams (New Brunswick, N.J.: Rutgers University Press, 1994), 123.

30. Ibid.

31. Žižek's discussion of the "act" here again proves fruitful. He finds, in Rossellini's description of the end of his film *Stromboli,* a description of the act: "There is a *turning point* in every human experience in life. . . . My finales are turning points. Then it begins again—but as for what it is that begins, I don't know" (Žižek, 43; italics mine). Thus the act summons a momentary symbolic destitution that transforms the symbolic field but with no control or concern for the shape of that outcome. The morph, as a punctuation between two cinematic regimes, functions precisely in this manner.

32. The first action hero film is generally taken to be *Rambo: First Blood, Part 2* (1985). Here I am positing *T2* as the last (or at least as the end of its era of unchallenged box office dominance).

33. See Yvonne Tasker's *Spectacular "Bodies": Gender, Genre and the Action Cinema* (London and New York: Routledge, 1993) for the prehistory and promise of the "action heroine." Although *Terminator 2* promised great changes for the role of women in action films, little has actually materialized. Since her return to action films in 1997 (with *Dante's Peak* and *Shadow Conspiracy*), Linda Hamilton has been given more stereotypically gendered roles, as have most other women

who have played visible roles in recent action films (e.g., Vivica A. Fox in *Independence Day* [1996], Maria Pitillo in *Godzilla* [1998]).

34. The demise of the action hero is arguable, at least, for the *white* action hero; *Passenger 57* (1992) and *Demolition Man* (1993) may also represent turning points along racial lines in this genre (although the emergence of the gangsta film complicates this transformation). *Passenger 57* is a thinly veiled reworking of the action heroics of *Die Hard* with Wesley Snipes as the action hero. *Demolition Man* portrays a struggle between Stallone and Wesley Snipes as equal rivals—although its pathologization of Snipes's character perpetuates a certain good guys wear/are white, bad guys wear/are black mentality. A similar transformation may be afoot with the emergence of Jackie Chan as a new international superstar (perhaps to be followed by a host of other Asian superstars—Chow Yun Fat, Jet Li, Sonny Chiba, etc.—whose international renown is also on the rise).

35. Patrick's string of featured appearances in the two years after *T2* presents a quite undistinguished litany of relatively unremarkable films; clearly *T2* does not represent his speedy (human) ascension to the action hero pantheon. In fact, his only roles in big-budget features in those years were cameo reprisals of the T-1000 in *Wayne's World* (1992) and *Last Action Hero* (1993).

36. Albert Liu, "Theses on the Metalmorph," *Lusitania* 1, no. 4 (1992): 131–42.

37. "Digital domain," which has become a fairly common shorthand for referring to the realm of computergraphic effects, happens also to be the name of one of the two leading digital effects houses, founded, not surprisingly, by James Cameron.

38. Certainly, Hollywood still believes in the selling power of stars—attested to by the recent proliferation of $20 million paychecks. However, such paychecks (and the endorsement of the star system that they suggest) are now justified solely by the international and video markets. This shift in star power is extremely important as a symptom of a shift in cultural logic. Both apparatus theory and the discourse on stars have traditionally relied on a notion of the larger-than-life import of the star as ego ideal, but clearly this model has been substantially transformed. Stallone's recent string of flops (*The Specialist* [1994], *Assassins* [1995], *Judge Dredd* [1995], and *Daylight* [1996]), and yet his continued $20 million deals, is especially indicative of this decreasing importance of the domestic box office in determining the financial fate of the action hero film.

39. In a hilarious indication that the *Jurassic Park* model continues to develop in these films, *The Daily Show*'s weekend box office wrap-up (originally aired 1 June 1998) summarized *Godzilla*'s lack of a credible human star by saying that "Matthew Broderick will make you long for the excitement that is Jeff Goldblum."

40. This list is limited to 1995 simply to avoid the necessity of continually up-

dating it to include the latest blockbusters. However, the trends that I have cited here have continued in the interim in films such as *Independence Day, The Lost World: Jurassic Park* (perhaps in even more extreme form than its predecessor), and *Godzilla.*

41. Indeed, as Vivian Sobchack pointed out to me, the very term "aftermath" that I use to describe the current moment contains the tumult and contradiction that I am describing. That is, at one level, the posthuman is "after math"—after the apparent elision of the laborious process of bodily calculation in a move to seeming effortlessness—and on another level, the posthuman is truly all about "math," the binary sequences and algorithmic calculations of digital technologies.

III. Morphing, Identity, and Spectatorship

9

Morphing Saint Sebastian Masochism and Masculinity in *Forrest Gump*

JOSEBA GABILONDO

This essay will examine the relationship between representation, morphing, and masculinity, focusing on the film *Forrest Gump* (1994).[1] My aim is to lead discussions of morphing away from merely technological concerns and toward its historical effects in the representation of subjectivity and ideology. More specifically, morphing will be studied as one of the key new technologies mobilized to renegotiate and relegitimize masculinity as Hollywood's central subject of representation. My contention is that throughout the 1990s, morphing becomes instrumental in articulating a new masculinity that is masochist, lacks agency, and yet is capable of reappropriating and reproducing difference for its own legitimation. At the same time, my specific analysis of *Forrest Gump* will foreground the fact that masculinity's new legitimation, through morphing, is highly contradictory and presents a utopian component that must be rescued for critical uses other than masculinity's legitimation.

In this context, my choice of *Forrest Gump* may seem perverse in that compared to more obvious choices, it occupies a peripheral place in relation to both morphing and masculinity. However, I hope to demonstrate that such peripheral positioning offers a privileged site from which both morphing and masculinity can be rethought and yet avoid the celebratory and fetishistic effect that new technologies and pumped-up masculinities usually have on criticism.

Morphing, Hyperreality, and Suture

It is important, first of all, to examine morphing in the very technical ways in which it affects representation. Ultimately, such technical analysis will make room for an examination of masculinity considered not simply as a general formation, accidentally related to morphing, but as one articulated by morphing in its very concrete and historical representations. To date, it could be argued that the three most outstanding applications of computers to film are digital compositing, warping, and morphing. *Who Framed Roger Rabbit?* (1988), *Jurassic Park* (1993), and *Forrest Gump* (1996) are the best examples of the first;[2] *The Mask* (1994), *Twister* (1996), and *The Nutty Professor* (1996), of the second; and *Terminator 2: Judgment Day* (1991), of the third. Although entire films can nowadays be generated by computers, their low resolution (in comparison with traditional film) has usually not made that desirable.

These three new applications of computers to film raise two historical paradoxes. The first concerns innovation. The new filmic applications that function most like older ones seem to promote technological innovation most effectively. That is, films that use digital compositing to *seamlessly integrate* computer-generated dinosaurs or aliens with live action (*Jurassic Park, Independence Day* [1996]), or American history with new characters *(Forrest Gump),* deliver the most powerful message in terms of innovation because of their perceived "realism." The films primarily based on the *spectacle* of morphing or warping *(Terminator 2: Judgment Day, The Mask)* become more identified with fantasy than "realism." Consequently, the latter films are contemplated in terms of their unique and isolated technological novelties rather than as representing general innovation. Coverage by the media as well as by other institutions is different in each case: whereas *Jurassic Park, Forrest Gump,* and *Independence Day* become widely discussed events, films such as *Terminator 2: Judgment Day* are deemed significant only for their technological wizardry.

The second paradox concerns the use of new computergraphic effects. The most visually stunning technological effects are the least used by film, although they remain the most impressive. Films such as *Jurassic Park* and *Independence Day,* as well as the *Star Wars* trilogy, primarily use digital compositing, not warping or morphing; indeed, the ratio of compositing to morphing or warping seems very disproportionate and favors compositing.[3] One could summarize and historicize both paradoxes by saying that the new representational technologies become most innovative and powerful at the point at which they are able to present the newest filmic realities while remaining most mimetic and referential, that is, realistic. Somehow, as if it were an asymptotic equation, there is an equilibrium to be met. Warping and morphing defy traditional realism (referentiality and mimesis) of film technology, whereas films that are too realistic (too referential and mimetic) do not display the technological innovation that has moved beyond the realm of the photographic.

I would like to propose that rather than first looking at morphing and warping separately, we begin by studying computer or digital image production as a whole and then move on to think about morphing and warping in relation to digital compositing. This approach will help us to understand why morphing and warping best capture the innovative nature of filmic computer graphics and yet remain marginal while digital image compositing, although more traditional in effect, proves more successful in the new cinematic economy brought about by all three technologies together. My goal is to explain the historical importance of these technologies as both a system or economy of representation and a complex group of different and specific technological applications, each with its own effects.

Recalling André Bazin's approach to realism and film technology might be helpful to our understanding of the historical importance of new computergraphic technologies as a whole system and economy of representation. Bazin long ago understood, although in quite Platonic terms, that the "myth of total cinema" historically preceded its technological materialization. The ultimate goal of this total cinema was an "integral realism":

> The guiding myth, then, inspiring the invention of cinema, is the accomplishment of that which dominated in a more or less vague fashion all the techniques of the mechanical reproduction of reality in the nineteenth century, from photography to the phonograph, namely an *integral realism, a recreation of the world in its own image, an image unburdened by the freedom of interpretation of the artist or the irreversibility of time.* If cinema in its cradle lacked all the attributes of the cinema to come, it was with reluctance and because its fairy guardians were unable to provide them however much they would have liked to.[4]

I have no doubts that the monomaniacal personality that Bazin detected in Edison and Lumière, both central to the development of "total cinema," also applies to the cases of James Cameron, George Lucas, and Steven Spielberg. However, in the contemporary case, computer technology has pushed film far beyond "integral realism"—unless the term is understood, counter to Bazin, as "hyperrealism."[5] In other words, and taking Bazin's theories to their latest consequence, computer technology has pushed "total cinema" into a new order of reality—the hyperreal—in which the representation of the fantastic does not delegitimize realism but rather legitimizes it as a stronger form of realism. In short, it is important to acknowledge that, ironically enough, Bazin already foresaw the logic of hyperreality when he suggested: "Every new development added to the cinema must, paradoxically, take it nearer and nearer to its origins. In short, cinema has not yet been invented!"[6] Through computer technology, cinema has pushed its origin back to premodernity, to medieval or older times in which the line between the fantastic and the real was not yet clearly drawn.

If the discussion of hyperreality is applied to the case of *Forrest Gump*, one can establish the way it indeed constitutes the general representational economy of Hollywood and is not simply an exceptional technological curiosity constructed to promote and exploit a particular technological innovation (such as warping or morphing). It is important to understand that in *Forrest Gump*, even though only digital compositing is used, the resulting representation is no longer realistic but hyperrealistic. That is, the compositing of past and present, accomplished through digital filters in *Forrest Gump*, points not only to the high level of realism achieved by the film but also to the technological nature, the hyperreality, of this realism.

In digital compositings, such as *Forrest Gump*'s, an actor's representation is inserted into a background that the actor cannot access physically and historically (i.e., Tom Hanks into footage of John F. Kennedy, or John F. Kennedy's image into new footage of Tom Hanks).[7] Neither the actor's representation nor the background changes in their mimetic shape and diegetic value. For mimetic and diegetic purposes, there seems to be no hyperreality, but rather "plain old reality." Nonetheless this form of digital compositing constitutes hyperreality—its almost invisible threshold—because the interface between actor and background already changes the filmic field or the mise-en-scène of representation altogether. It is important to remember that because of the historical difference in the source of both images, the compositing can no longer be accomplished optically ("real" cinematic compositing) but only digitally (hyperreal compositing). If hyperreality is clearly present in the most traditional and invisible application of new technology, such as compositing, one could conclude that hyperreality is indeed the filmic economy that now constitutes and regulates filmmaking as a whole in Hollywood.[8] Thus hyperreality is not simply a technological effect of certain science fiction films. In other words, one could claim that films such as *Forrest Gump* do not constitute the mimetic referential standard from which films based on morphing or warping are seen as the exception. Rather, morphing or warping, as much as digital compositing, are constitutive parts of the same hyperreal economy.[9] Hyperreality establishes the representational continuity between digital compositing, warping, and morphing.

This representational continuity is also established by the fact that in digital compositing, representations become hyperreal through morphing. As I will argue, digital compositing morphs real images into hyperreal representation. Here the meaning of "morphing" no longer is metaphorical in the sense that it is used loosely out of its normative context. In what is traditionally called morphing (i.e., figural morphing), the change from one shape into another is given in *time,* and thus the viewer can follow the change as it unfolds in *space.* In hyperreal or digital compositing, however, the change takes place in space. But on the one hand, the viewer can perceive the real images composited from different filmic sources or realities: Tom Hank's image from present film footage, John F. Kennedy's image from historical 16 mm film footage. But on the other, the viewer can also view the way these images are morphed into a single hyperreal image: Forrest Gump meets President John F. Kennedy in the new footage of the film *Forrest Gump.* The viewer is aware of the *double temporality* of the image: its real and separate origins as well as the resulting morphed single present, a hyperreal present. However, the viewer does not cancel the difference between both temporalities, no matter how little or imperceptible their threshold is. The viewer can see the original "realities" and their morphing, the real and its morphed hyperreal difference, as a simultaneous process. Both moments, the original real and the resulting hyperreal, take place simultaneously while their temporal difference is preserved. In this sense, digital compositing is morphing: it captures simultaneously in space the two images' hyperreal change in time.

Furthermore, and as the following examples will illustrate, even at the periphery of digital compositing, strict or figural morphing has to be used to digitally suture the composited images and make them fully mimetic and referential, that is, realistic. As the director of *Forrest Gump,* Robert Zemeckis, explained in an interview with *Premiere* magazine, the technical team composited the images in which Forrest Gump meets three presidents of the United States in two complementary ways.[10] Sometimes they inserted the image of a given president in new footage and then digitally filtered the footage to make it look contemporary to the president's footage. Other times, however, they inserted the image of Forrest Gump/Tom Hanks in old footage and then digitally filtered his image to achieve a historical continuity between images. It is important to emphasize that in order to fuse both images, and make the compositing perfect, several details in the images had to be morphed, technically speaking. First, once the image compositing was achieved, the edges between the inserted image and the background had to be digitally treated to make the historical transition smooth and believable. Thus the edges of the two composited images needed to be morphed to eliminate their disparate real origins and their different traces. Second, the mouths of the presidents had to be morphed to make their lip movements correspond to the dialogue of the film script. Third, certain actions such as the handshake between John F. Kennedy and Forrest Gump also had to be morphed so that it looked as if they were actually taking place. Finally, the only full figural morphing in the film was used to show Tom Hanks becoming one of the Klansmen in *The Birth of a Nation.* Tom Hanks's footage, filtered to look old, slowly morphs into actual footage of a Klansman in *The Birth of a Nation* in such a way that for a few seconds, the viewer is watching *The Birth of a Nation.*[11] As these examples show, morphing is nearly always a constitutive element of digital compositing. Thus in the end, hyperreal compositing is itself a highly mimetic and referential form of morphing.

Once digital compositing is redefined as a form of morphing, it becomes paramount to establish its relationship with warping and strict figural morphing. First of all, it is important to point out that in my view, there is not any qualitative difference between warping and figural morphing; this despite the fact that every single manual on the topic contains a technical definition of "(figural) morphing" as derivative of, but qualitatively separate from, "warping." Warping involves only one image and figural morphing two, hence their assumed difference. For instance, Crane gives the following definition: "Image warping is the act of resampling an input image according to specific rules. . . . Morphing is a term mutated from the word metamorphosis, consisting of image warping and cross-dissolving two images."[12] A manual by Watkins, Sadun, and Marenka follows this general definition as well: warping one image, figural morphing two.[13] However, and contrary to the technical literature, I would like to claim that in representational and historical terms, it makes more sense to stress the continuity between both processes and to consider warping a derivation of figural morphing.

Although no specialist has dwelled on the problem, the distinction made between the number of images involved in warping (one) and figural morphing (two) is less technical than cultural. In the case of warping, there is always a second image present at the end of the process of warping: the distorted image of the original. Indeed, the idea of a second image is *necessary* to the concept of warping. I will refer to a simple example. When in *Terminator 2: Judgment Day*, the Terminator T-1000, morphed as a policeman, is thrown against the wall, instead of turning around, the figure warps/morphs so as to reproduce a new face and front from its back. Are the original and final image the same or different? Are they one or two images? Is this morphing or warping? The continuity or discontinuity between the original and the resulting distorted image is established according to criteria of identity. If their identity is ascertained, it follows that both are the same image, one single image, and the process gets called warping. However, if the identity of the distorted image can no longer be traced to the original, it is assumed that they are two different images, and the process is called "morphing." Distinguishing the identity or difference between the original and the distorted or final image requires the a priori distinction of both images. Consequently warping is a type of figural morphing where the identity between the original and final image can be established, and the resulting change is called "distortion." In short, *warping derives from figural morphing*. Ultimately there is no radical difference between both techniques, only degrees of difference. Difference and identity are not technological issues but cultural; or as Heidegger would have it, "the essence of technology is nothing technological."[14]

In sum, if warping is a form of figural morphing, and digital compositing is a basic form of morphing as well, one could conclude *that morphing constitutes the basic economy of hyperreality*. This assertion resolves the two paradoxes mentioned previously. That is, over—and under—all, morphing is the most innovative and widely used form of digital representation. All differences in the processes are a matter of degree.[15] Thus to keep the historical value of the differences between compositing, warping, and figural morphing while stressing their continuity, I will call digital compositing "zero degree of morphing" or "low morphing," warping "medium morphing," and figural morphing "high morphing." The resulting hierarchy attempts to stress the primary underlying importance of compositing in morphing while acknowledging the high degree of morphing involved in figural morphing. This hierarchy can be illustrated thus:

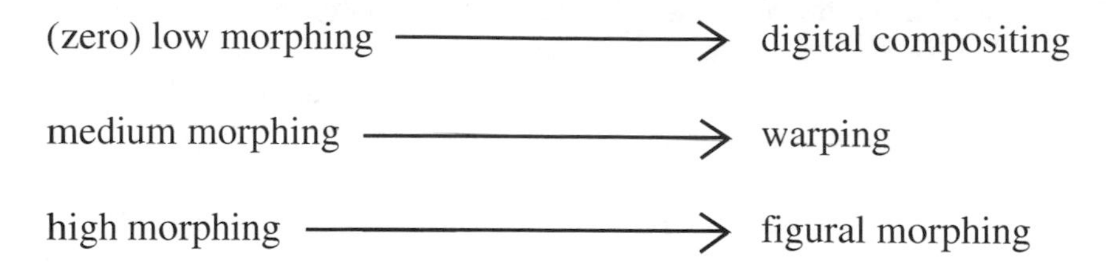

Figure 9.1 Morphing economy and hierarchy. Courtesy of the author.

To explain the articulation between these three degrees of morphing (understood in its new sense of constituting a general representational economy of hyperreality), it is worth examining the mimetic and referential effects of each. In general, the anxieties generated about the "mimetic and real" effects of morphing lead the audience to question all digital representation as technological and "artificial." However, each technique delivers a different degree of "mimesis and artificiality." In digital compositing, the represented elements keep their full mimetic value. Because composite morphing is simultaneous and visible in time, its mimetic representation is most effective, and the realist anxiety of the viewer is lowest. Warping and figural morphing, because of their temporal nature, undermine the mimetic value of representation more than compositing and, as a result, stress their antimimetic artificiality, and irreality. The moment the original image is warped or morphed, its mimetic value is distorted as well, and thus mimesis is jeopardized altogether. Only at the end of the process, when the final image is recognized as a realistic representation, is its mimetic value restored. However, in warping, because the identity between the original and final images is upheld, the mimetic value of the process is undermined less than in figural morphing. In the latter, the resulting identity difference between the original and final images creates higher mimetic and realistic anxieties. Thus one could establish a mimetic hierarchy between the three techniques: the higher the degree of morphing involved, the less mimetic the process becomes. In other words, the ratio between morphing and mimesis is inverse.

Once the relationship between mimesis and morphing is established, one could theorize the way morphing contributes to "suture" and maintains dominant ideology. Stephen Heath states: "Suture . . . names the dual process of multiplication and projection, the conjunction of the spectator as subject with the film . . . which conjunction is always the terrain of any specific ideological operation of a film."[16] In the context of Hollywood, mimetic realism is the main discursive strategy by which film creates suture. Thus one could summarize the suturing effects of morphing by saying that the lower the degree of morphing, the higher the mimesis/suture and the ensuing ideological effect of realism achieved by film. This pseudo-mathematical equation takes us beyond the depoliticizing effect created by Baudrillard's theory of the simulacrum as installing a homogenizing hyperreal economy with no difference and politics. Furthermore, even from a fiscal point of view, high suture—achieved through digital compositing—is the most labor-intensive work. As Ron Magid summarizes in relation to *Forrest Gump:*

> No film in history has used so many effects to mimic what we think of as reality. Intent on wringing the most emotion from every sequence, Zemeckis ordered his troops to turn calm seas turbulent, make clear skies cloudy, and fill empty space with moving spectators. "*This project is four times the size of* Jurassic Park *in terms of complexity and the size of the crew,*" says Rosenbaum, who, along with computergraphics supervisor George Murphy and computergraphics sequence supervisor John Schlag, implemented Zemeckis's and Ralston's vision of *Forrest Gump.* "*People won't realize that most of the effects in the film were designed to enhance reality.* I find that more enjoyable than working on a fantasy film."[17]

Finally, the relation between mimesis and suture would in turn explain the previously mentioned second paradox: the most visually stunning computergraphic effects are the least used by hyperreal films. The low degree of suture present in high-degree morphing would account for the scarcity of films centered around such effects. It is harder, one could argue, to hold the viewer's identification and desire.[18] Indeed, the highest degree of morphing would create abstract film. At the moment the technique fully represents itself, it also disappears as re-presentation. Thus films containing high degrees of figural morphing are understood as science fiction or fantasy, and ultimately in their higher degree are "othered" as irreal.[19]

Masculinity's Representational Surplus Value and Its Utopia

The foregoing theoretical approach to morphing is productive only insofar as it is understood as a methodological moment in which morphing is isolated for the sake of its systemic description. Historically speaking, however, morphing—to any degree—is used to constitute narrative and cultural meaning. However, now that we have a unified theory of morphing encompassing digital compositing, warping, and figural morphing, we are in a better position to analyze the historical repercussions of morphing.

As morphing consolidates as a technology and technique throughout the late 1980s and early 1990s, it becomes a constitutive part of the cultural history and politics of that transitional period and, more specifically, becomes key in the formation of different political struggles centered on representation. That is, if morphing's cultural articulation is analyzed, rather than merely its semiosis, one must acknowledge that morphing has become a major tool for the representation and legitimation of white heterosexual U.S. masculinity (henceforth Masculinity). Here, I will leave aside the different uses of morphing in other cultural formations such as globality and postcoloniality, consumerism, and so forth. Instead, I will concentrate on Masculinity, although it will be situated as an integral part of these other ideological formations.

Susan Jeffords analyzes how the hard-body Masculinity of the 1980s experienced a softening at the end of the decade and discusses this shift in relation to race:

> As has historically been the case in dominant U.S. cultures, masculinity is defined in and through the white male body. . . . Action films of the 1980s reinforce these assumptions in their characterizations of heroism, individualism, and bodily integrity as centered in the white body. And though 1991 films repudiate many of the characteristics of that body—its violence, its isolation, its lack of emotion, and its presence—they do not challenge . . . the "special" figuration that body demands. If, these films suggest, there is a body that has been betrayed, victimized, burdened by the society that surrounds it . . . it is . . . the body of the white man who is suffering because he has been unloved.[20]

If anything, then, the turn taken by Masculinity in the 1990s can be defined by its narcissism. That is, the wounded and introverted Masculinity of the 1990s has not renounced its eighties hard body; rather, it has problematized it. To the hard body it inherited from the 1980s, 1990s Masculinity has added a long inventory of bodily paraphernalia to fetishistically study and resignify its self-involvement. On a bodily level, one could list the following: well-toned muscles (no longer gigantic but very noticeable nevertheless), long hair, hair coloring, sideburns, tanning, tattoos, piercing, and branding. At the level of accessories, a nonexhaustive list would include earrings, colorful psychedelic sixties and seventies revival clothing, leather, and baggy clothing. Furthermore, even Masculine sexuality has become narcissistic. Consequently the borders between heterosexuality and queerness have been narcissistically problematized by Masculinity. Its new exposure to "queerness" is a well-negotiated move: Masculinity explores and redefines itself as (post-) heterosexual (albeit not queer) through its encounter with the queer body (male and female). Indeed, the rise of mainstream queer characters in films and television programs, which could be symbolically dated to the release of *The Crying Game* (1992), is part of this narcissistic phase of heterosexual Masculinity. Finally, as Jeffords explains, even whiteness becomes a site of narcissistic racial relegitimation for Masculinity.

However, in this new general narcissistic turn to a "souci de soi,"[21] there is one element that cannot wholly be appropriated by Masculinity: reproduction. Jeffords points out that the problem of (self) reproduction is at the core of Masculinity's formation. Furthermore, she notes that Masculinity's (self) reproductive strategies shift from the 1980s to the 1990s through its deployment of "inversion rather than duplication."[22] That is, Masculinity becomes preoccupied with narcissistic or "inverted" reproduction. As a result of this self-centered preoccupation with reproduction, Masculinity reclaims its fatherhood and also attempts to claim motherhood as well as part of its domain. As Jeffords argues about *Terminator 2: Judgment Day:* "By 'giving' John Connor his life, the Terminator takes, in effect, Sarah Connor's place as his mother. In one of the film's most astounding inversions, the Terminator can now be said to have given birth to the future of the human race."[23]

In this respect, *Forrest Gump* can also be seen as paradigmatic of this negotiation, although it operates both more subtly and more pervasively than do its more flamboyant counterparts such as *Terminator 2.* The key to *Forrest Gump*'s "subtlety" is morphing. Indeed, morphing has become one of the most interesting ways in which Masculinity has managed to "reproduce" itself narcissistically and fashion *a new postfeminist misogyny based on nonbiological reproduction.* In the case of *Forrest Gump,* Masculinity's repositioning within a morphed history allows for its narcissistic reproduction and thus the reproduction of history. Certainly, at the narrative level, *Forrest Gump* displays the narcissistic and reproduction-obsessed character of Masculinity in the 1990s. As Thomas Byers explains:

> Sweet, innocent, polite, and chivalric child-man; devoted son to his mother; brother of the nice Black male (Bubba's bubba); best friend and savior of the disabled veteran; patient, all-forgiving spouse and hospice nurse to his wayward Jenny; and nurturing Mr. Mom to his own son, Forrest represents a liberal myth (in Barthes's sense in "Myth Today" [143]) of the boomer as the "new man," egalitarian, sympathetic to the marginalized, and in touch with his "feminine side. . . ." At the same time Forrest is, by turns, an All-American football star, a Medal-of-Honor-winning war hero, a wildly successful entrepreneur, a spiritual leader held in awe and reverence, and a fertile and wise father, thereby also living up to the fantasies of a traditional masculinity. He combines within himself the virtues of both the strong father and the saintly mother of the fifties.[24]

However, it is important to recognize that the narrative history of *Forrest Gump* does not explain why Forrest Gump, a passive and hardly macho character despite all his accomplishments, becomes an icon for Masculinity. This iconicity, I would argue, is accomplished through morphing, and Forrest's meeting of the great figures of contemporary American history—with the symptomatic exception of Martin Luther King Jr., as pointed out by Byers.[25] Only morphing allows Forrest Gump to reproduce himself as a historical and national subject, to stand for Masculinity. The morphed Forrest Gump becomes a technological mirror in which Masculinity can contemplate itself reproduced in history. That is, historical morphing warrants reproductive powers to Masculinity. Thanks to technology, Masculinity does not reproduce babies but rather reproduces entire nations and histories. Hence the superhero status of Forrest Gump: his morphing reproductive powers transcend those of the female body and become, if anything, more motherly than any woman's.[26] This need for morphed reproduction is part of a larger representational strategy to rescue Masculinity, legitimize it, and endow it with a higher representational value. Morphed representation and reproduction have a surplus value, both technologically and historically.[27]

Analyzing Masculinity in the 1980s, particularly in relation to science fiction films, Vivian Sobchack and Jeffords both point out that time traveling allows the hero to father himself and thus appropriate reproduction for himself—a "fictional move" that, as Sobchack indicates, neither the horror nor family melodrama genres could have accomplished; thus the importance of SF as genre in the 1980s.[28] However, the hero has to live in two different times to accomplish self-fatherhood: Masculinity splits into two Masculinities discontinuous in time: it doubles itself. After morphing is introduced, however, the narcissistic doubling and misogynistic reproductive ambitions of Masculinity are accomplished in the surplus of its own time, simultaneously. If Jeffords notes that Masculinity's reproduction shifts in the 1990s *from* doubling *to* inversion, it is necessary to amend her analysis by saying that 1990s Masculinity reproduces itself by *both* doubling (two Masculinities) *and* inverting itself (mirroring itself narcissistically in its own morph).

In the most accomplished example of the early 1990s, *Terminator 2,* it is important to emphasize that the T-1000 model—which morphs into liquid metal—is a double of the original Terminator (played by Schwarzenegger), who cannot morph. Here again Masculinity doubles itself. Both Terminators, in their duplicity, are allowed to appropriate and reproduce *all* the subject positions of the film: from the new loving fatherly and motherly Masculinity of Schwarzenegger's original Terminator to the sadistic Masculinity of the T-1000 (played most often by Robert Patrick).[29] Masculinity is embodied not only by Schwarzenegger, as Jeffords would have it, but also by the second morphing Terminator (T-1000), which in turn is allowed to carry out all the sadism proper of 1980s Masculinity. This Masculinity is as reproductive as any mother and yet possesses the surplus value of *total* reproduction: it can even reproduce reality, although it needs first to destroy that which it reproduces afterward.

In most of the 1990s films, from *The Mask* to *The Nutty Professor,* the narcissistic doubling of Masculinity is carried out by morphing. This doubling always announces that Masculinity's reproductive powers can transcend women's because men now have the surplus value conferred by morphing. At the same time, this surplus value allows the doubled protagonists of these films simultaneously to experience, on the one side, the normative Masculinity of the 1990s (gentle, sensitive, caring) and, on the other, the older Masculine norm of the 1980s (hard, violent, sadistic). In both *The Mask* and *The Nutty Professor,* the morphed sadistic Masculinity is eliminated at the end, but only after the unmorphed or "real" Masculinity learns from the morphed one how to reappropriate sadism in a socially acceptable way.

Needless to say, in these two cases, the marked ethnicity of both masculinities moves them to the periphery of Masculinity proper.[30] That is, ethnically and racially marked masculinity is contained within the comedy genre, where it is turned into a carnivalesque and popular (nonofficial) countercelebration of hegemonic Masculinity. In *The Mask,* every time Jim Carrey's character is morphed by the mask, he wears ethnically coded costumes, most of them Latino. Indeed, the zoot suit is his prevalent costume, and Carrey sings a Pachuco song. One could assume that here, in a comedy, Masculinity is allowed to experience any form of difference, including race and ethnicity, through the power of its "fatherly reproduction": morphing. However, at the end of *The Mask,* the morphed Pachuco is the one annihilated. An opposite move is carried out in *The Nutty Professor,* also to keep racial difference in check. Eddie Murphy manages to represent, through reversed morphing, a misogynist form of African American masculinity that nevertheless possesses all the critical potentiality of the trickster. The film tells the story of an overweight, self-conscious, asexual scientist ("Nutty Professor"/morphed Murphy) who manages to find an unstable chemical formula to turn temporarily into his alter ego: a slim, witty, uninhibited, testosterone-driven lover ("Buddy Love"/normal Murphy). Only by morphing and annihilating the latter form of masculinity can the Murphy/Nutty Professor be allowed back into "normal and respectable" white society. Here, he is accepted as the "monstrous" overweight and morphed Eddie Murphy with no trickster abilities. Through morphing, then, critical masculinity is castrated. Both these films disclose (white) Masculinity's ultimate ideological fantasy: morphed reproduction can transcend not only gender but also race and ethnicity; *Masculinity transcends social difference because it can reproduce difference through morphing.*

Finally, it is important to direct attention to the fact that the narcissistic reappropriation and self-reproduction through morphing also affects representation altogether. In *Forrest Gump,* for example, history is reduced to representation susceptible to Masculinity's reproduction.[31] In this way, *Forrest Gump* not only draws a Ku Klux Klan scene from *The Birth of a Nation* but also entails the filmic genealogy inaugurated by Griffith's 1915 film, a genealogy that shows film no longer as simple entertainment, socially contained, but as a technological alteration and rewriting of social and historical imagination.

If Masculinity's reproductive morphing is analyzed at a psychoanalytic level, the inherent sadomasochism that historically articulates

Masculinity is here present in a very specific formation. Laura Mulvey explained long ago the explicit sadism involved in the formation of Masculinity in regards to visual pleasure and the images of women in film.[32] Subsequently Steve Neale has pointed out, in one of the foundational articles on masculinity and film, that masculinity is also articulated by a high degree of sadomasochism among men in regards to their visual pleasure while watching film images of males: "The repression of any explicit avowal of eroticism in the act of looking at the male seems structurally linked to a narrative content marked by sado-masochistic phantasies and scenes."[33] The morphed and reproductive Masculinity of the 1990s shows a new arrangement of this sadomasochistic economy: masochism becomes *avowed* by traditional filmic means (narrative and camera) while sadism is *disavowed* through morphing.

Forrest Gump's encounter with history evidences such a new arrangement. Masochism, because it is avowed both by the narrative and the classical use of the camera, is clearly marked throughout the film. First, in relation to the narrative, Forrest Gump is always presented as *suffering* history; he never acts out history, but instead history befalls him. He has no choice but masochistically to endure its consequences. Then masochism is avowed in *Forrest Gump,* regarding the point of view articulated by the camera, in ways that depart from previous historical representations of Masculinity. It is important to remember that in most action films of the 1980s (*Lethal Weapon* [1987], *Die Hard* [1988], et al.), there is one scene in which the camera *relinquishes* the Masculine point of view. Instead it presents the hero's defenseless and wounded image for the contemplation of the viewer. It is a masochistic scene in which Masculinity presents itself as defenseless in order to gain the love, sympathy, and identification of the viewer—even at the risk of homoeroticizing Masculinity. Once this masochistic identification is obtained, however, Masculinity is rearticulated and now, with the viewer's legitimation, sets out to pursue its sadist objectives. In the case of *Forrest Gump,* the masochist point of view is prevalent throughout the film. That is, for most of the film, Forrest Gump is contemplated by the viewer: the camera does not construct a classical sadistic Masculine point of view aligned with Masculinity's agency. By Hollywood standards, the film presents an unusual number of long takes with a shot-in-depth structure as well as only a few shot/reverse-shot sequences.[34] Thus identification is achieved not through Masculine desire but rather by the *suspension* of Masculine desire proper to masochism.[35]

The film's disavowal of sadism unfolds through morphing, rather than through the camera, as all the different characters who meet Forrest Gump also meet a tragic destiny, and he is the only one who survives. That is, Forrest Gump's morphed encounter with history is sanctioned and marked by violence, castration, and death. Forrest is always morphed into encounters with historically important people who are victims of violence, castration, or unnatural death. By the end of the film, nearly all the historical characters who meet Forrest Gump die: Elvis, John and Robert Kennedy, George Wallace, and John Lennon. Lyndon B. Johnson seems to be the only president not affected (although he will not run for a second term). Richard Nixon is shown resigning from the cabinet, and Gerald Ford and Ronald Reagan are shown only through footage of assassination attempts on their lives.[36] Furthermore, the rest of the film's characters die as well: Forrest's mother, Bubba, and Jenny.

As the characters disappear, every historical difference they mark is erased as well. Because of the "buddy film" structure of *Forrest Gump,* the death of Bubba becomes especially important because it opens up the possibility for a postracial history. Furthermore, after Gump's encounter with the last of the three presidents he meets, Nixon, this history also becomes *postnational.* In other words, other abject forms of masculinity are denied national and historical transcendence; consequently they disappear as history unfolds rather than morphing through it. Only Forrest Gump is able to morph throughout history and beyond. Those masculinities marked by difference are thus mobilized and dismissed at will by Masculinity. Only Forrest's white war buddy, Lieutenant Dan Taylor, survives, although he is clearly marked by violence: his legs are amputated. Ironically, but aptly, the actor's real and full legs are digitally castrated through morphing. Indeed, this morphed castration can be read allegorically as the transition of 1980s Masculinity into the masculinity of the 1990s. Here again Masculinity is doubled and inverted through morphing. Lieutenant Dan represents the Masculinity that does not disavow sadism; hence his morphed castration. He thus becomes a shadow of the angry sadist man that the film disavows. In short, morphing allows a retelling of the national discourse of Masculinity. (It is no coincidence that the favorite figure for morphing demos is usually the founding president of the United States.)

At the end, Masculinity prevails over history, nation, and race. American history and Forrest Gump's biography coincide, and the flashback narrative stops when Forrest leaves the bench from which he has told his story to the other passengers awaiting their buses and he walks to Jenny's apartment. Over the course of the film, American history becomes Forrest Gump's history and, from this moment on, also his personal life. At the end, the posthistorical subject position enacted by Forrest Gump moves to the domestic level and there again takes center stage through a new postfeminist and neomisogynist refashioning.

If any character in the film represents and condenses actual history, this character is Jenny, and she is a woman. Jenny keeps fleeing and abandoning Forrest, thus shaping his masochism at the narrative level. Also showing up again and again, she marks history's return into Forrest Gump's life. Jenny represents the American history of the 1960s counterculture and the changes it brought to conservative 1950s America. As Byers argues: "Jenny . . . is the figure of just about everything the New Right means by the counterculture: she is a folksinging bohemian, a 'loose' woman . . . an acid-dropping flower-child who hitches a ride in an old VW to San Francisco for the summer of love, an antiwar activist and lover of a radical leader, a disco-dancing cocaine addict, an HIV-positive single mother."[37] Her death from AIDS signifies the obliteration of history and biological motherhood, and its subsequent displacement and reappropriation by Masculinity—and the final doubling of Masculinity in the biological inversion of Forrest Gump and his son. There is no longer need to morph Masculinity.

At the end of the film, Forrest Gump becomes the new postnational, postracial, postfeminist subject embodying 1990s Masculinity. However, this final and general disavowal of difference also captures the most interesting contradiction, politically speaking, of *Forrest Gump:* Masculinity needs to renounce not only its agency but also its history to gain legitimation and overcome its problems with race, gender, and reproduction. This is the final contradiction that *Forrest Gump* cannot solve.

I have purposely delayed discussion of an extremely important morphing moment in the film: the feather carried by the wind at its beginning and its end, and the longest digital composite in film history.[38] Dostoyevsky understood a long time ago that to write the story of a saintly person—a beautiful soul—he had to portray a person who neither affected history nor was affected by it. In other words, Dostoyevsky needed to write the story of an idiot, and his novel *The Idiot* is indeed a reflection on the issue of historical agency and morality.[39] In *Forrest Gump,* the attempt to narcissistically recuperate a reproductive Masculinity, one that can get away with posthistorical misogyny and racism, ends up constituting a form of earthly male sanctity—beyond history—that nears idiocy. The feather at the beginning and end of the film hints at the ethereal and angelic nature of Forrest Gump.[40] By disavowing sadism and avowing masochism, Forrest Gump becomes the model of sanctity that contemporary Masculinity wants to embrace. Indeed, the masochist sanctity of the likes of Saint Sebastian becomes the model for Masculinity in the 1990s.[41] However, a sanctified Masculine masochism has a historical price to pay: a loss of agency that generates ahistorical idiocy as its teleology.[42]

One could read *Forrest Gump* as part of a new organization of Masculinity reminiscent of that effected by the bourgeoisie in the nineteenth century, which Kaja Silverman has theorized following J. C. Flugel's account of "The Great Masculine Renunciation of the Body."[43] Bourgeois masculinity's renunciation of the ancien régime's sartorial and exhibitionistic dressing entailed its transformation into a bodiless and scopophilic subjectivity that could then observe and control "objective reality" as its difference, a difference including women and colonial subjects. Bourgeois masculinity's gaze structured an epistemological order—objective reality—of which it became the ultimate real subject. If *Forrest Gump* is an index of 1990s Masculinity, one must conclude that Masculinity has again renounced something to preserve its hegemony: this time not its body but rather its epistemological regime—"reality."[44] Instead Masculinity has embraced the uncertainties of hyperreality and morphing, with all their subsequent possibilities for decentering and loss of hegemony; indeed, Masculinity's lack of historical agency is only part of this conflictive new position.[45] This new "Great Masculine Renunciation of Reality" and its ensuing "agency-less" repositioning in hyperreality is the overarching masochist structure defining Masculinity in the 1990s.

At the beginning of this essay, I painstakingly delineated the articulation between digital compositing, warping, and morphing because such analysis opens up the possibility of discussing the suturing effects of morphing in relation to Masculinity. *Forrest Gump*'s "Masculine Renunciation of Reality" makes room for a technical discussion of hyperreal suture.

If *Forrest Gump*'s low morphing makes room for such a discussion of suture, it is indeed because the renunciation of reality that Masculinity effects through low morphing seems natural, biologically real, as opposed to high-morphing films such as *Terminator 2, The Mask,* and *The Nutty Professor.* Because of the high morphing and low suture present in these latter films, the hyperreal Masculinity they articulate is destroyed at the end of the narrative. At the end of *Terminator 2,* for example, mother and son bid both Terminators farewell. Hence the audience renounces both morphing and hyperreality. Reality's effects return at the end of the film. The spectator is back in old, comforting reality with Linda Hamilton and her teenage son, John. Here the Great Renunciation of Reality remains dystopian or at least melancholic. *Forrest Gump,* however, does not renounce morphing and its hyperreal Masculinity. It is rather the opposite. Nonmorphed reality is renounced as fake, as nonreal, as "old history troubled with social difference." As history is banished in *Forrest Gump,* so is reality. Consequently *Forrest Gump*'s new "hyperreal reality" is embraced as "the only real, the only present reality." Old reality looks fake and unreal compared to *Forrest Gump*'s new "hyperreal reality," which, as presented and embodied by Masculinity, offers a subject position capable of "transcending" social difference.

At the end of *Forrest Gump,* the morphed father and his offspring bid farewell to the only real, nonmorphed character left in the film: the woman, the mother, the ultimate embodiment of history and difference. This is the high suture effected by *Forrest Gump*'s low morphing, one in front of which high-morphing films such as *Terminator 2* pale with ideological envy. In *Forrest Gump,* we gladly embrace hyperreality as our new reality, and in the process we accept Masculinity as its sole subject. Indeed, in the film, Forrest Gump, the subject who has overcome history, reality, and all forms of social difference, looks more real than any U.S. president—or historical event we have seen.

Low morphing allows this transition: it appears as real as "old reality" but captures its own changing historicity in a way that "old reality," represented through the conventions of realism, no longer does. Hyperreality changes historically because it now has a new subject: Masculinity. *Masculinity is the new historical subject that renounces reality and makes this renunciation the central event of history*—thus constructing a new hyperreal history, and reality. This is the utopian, if masochistic, moment of *Forrest Gump,* and it rests opposite to the dystopian scenarios of high-morphing films. Nevertheless, Masculinity's masochist renunciation of reality also represents a utopian possibility for other subject positionalities. *Forrest Gump* delivers the message that history no longer is a reality fixed by the truth of mimetic representation and its traditionally masculine subject. In short, history is no longer real, but hyperreal and open to change. Any subject can access history and rewrite it from its own position. History can be accessed and then morphed, reproduced, and changed in such a way that one's own position need no longer be marginal or peripheral—

although it does become hyperreal.[46] Thus, ironically enough, *Forrest Gump* announces the possibility of agency in history. Sobchack captures this historical contradiction in *Forrest Gump* when she notes: "While one can certainly argue its marking the dissolution and 'end' of history (as well as the responsibility for it), *Forrest Gump* can be argued also as marking (and dependent upon) a new and pervasive self-consciousness about individual and social existence as an 'historical subject.'"[47]

I am aware that this final utopian pronouncement about morphing presents the same problems of Donna Haraway's "Cyborg Manifesto": that is, the historical difficulty of utopically accounting for the relocation of subjectivity and political agency in hyperreality.[48] However, I believe that analyses such as this one point to the historical location where the political and cultural struggle for representation begins. It is clear by now that hyperreality is the new horizon for commodification in late capitalism or postindustrialism. However, masochistic films such as *Forrest Gump* indicate that the "subjects" of postindustrial commodification are no longer slavishly "subject" to Masculinity's hegemony. The space of historical indeterminacy between postindustrial commodification and Masculinity still remains a utopian possibility for positionalities other than Masculinity.

Notes

I would like to thank Susan Williamson from the Annenberg Center Library for her bibliographic help, Vivian Sobchack for her thorough editing, and Sharon Ullman for her endless insights and enthusiasm when discussing masculinity in mainstream film.

1. The film is based on the novel by Winston Groom, *Forrest Gump* (New York: Pocket Books, 1994).

2. Other remarkable examples are recent commercials in which contemporary footage about different products is composited with performances by deceased stars: for example, Fred Astaire's dance with a vacuum cleaner, and Natalie Cole's singing "Unforgettable" with her dead father, are part of this trend.

3. In other areas such as commercials, however, where the main function of the image is not "realistic" narrative representation but "shock" and "appeal," the ratio seems more proportionate.

4. André Bazin, "The Myth of Total Cinema," in *What Is Cinema?* trans. Hugh Gray (Berkeley: University of California Press, 1967), 21; italics mine.

5. Jean Baudrillard, *Simulations,* trans. Paul Foss, Paul Patton, and Philip Beitchman (New York: Semiotext[e], 1983), 2.

6. Bazin, 21.

7. See Paula Parisi, "Forrest Gump Gallops through Time," *American Cinematographer* 75, no. 10 (1994): 39–42.

8. Obviously "realistic" films with no digital output are present in Hollywood but have been pushed to the periphery of the blockbuster economy. Furthermore, the tendency to touch up pictures digitally (by deleting small details from the background or modifying certain features on an actor's face) is so widespread that the same category of "photographically real films" is endangered, if not already extinct.

9. The recent revival of cartoons as an adult form of entertainment is due in part to this consolidation of hyperreality.

10. Mimi Avins, "'Gump' Cut: When Tom Hanks Shows That Heart Is Better than Smart in 'Forrest Gump,' It's History in the Faking," *Premiere* 7, no. 12 (August 1994): 78.

11. Ron Magid, "ILM Breaks New Digital Ground for *Gump,*" *American Cinematographer* 75, no. 10 (1994): 49–50.

12. Randy Crane, *A Simplified Approach to Image Processing: Classical and Modern Techniques in C* (Upper Saddle River, N.J.: Prentice Hall, 1997), 203.

13. Christopher Watkins, Alberto Sadun, and Stephen Marenka, *Modern Image Processing: Warping, Morphing, and Classical Techniques* (Boston: Academic Press, 1993), 97. John C. Russ, in *The Image Processing Handbook,* 2d ed. (Boca Raton, Fla.: CRC Press, 1995), also begins the section on morphing by establishing the derivative nature of morphing from warping, even if in an indirect way: "Programs that can perform controlled warping according to mathematically defined relationships . . . are generally rather specialized. But an entire class of consumer-level programs has become available for performing image morphing based on a net of user-defined control points" (208).

14. Martin Heidegger, *The Question Concerning Technology and Other Essays,* trans. William Lovitt (New York: Harper and Row, 1977), 35.

15. The continuity of the different degrees of morphing can also be explained by the continuity in the software that supports it. Avid's Elastic Reality for Windows, for example, executes all the following tasks or functions: warping and morphing, shape-to-shape interface, assisted tracing tool, compositing and matte making, color correction, hierarchical 2-D animation and image transforms, and file format management. In other words, the software itself works based on the continuity of the different degrees of morphing. See "Avid Ships Elastic Reality 3.0

for Windows '95 and Windows NT," on-line, http://www.avid.com/news/press_releases/product_news/er3.0_nt.html, 13 May 1999.

16. Stephen Heath, "On Suture," *Questions of Cinema* (Bloomington: Indiana University Press, 1981), 109.

17. Magid, 45–46; italics mine.

18. The relationship between suture and desire elicited by morphing requires a separate discussion that would most likely focus on visual abjection.

19. I would like to bring attention here to the issue of editing. In traditional film theory, mise-en-scène and shot structure have always been differentiated and dealt with separately. Once digital technology is involved in image editing, the relation between digital compositing—low morphing—and editing becomes blurred, since image compositing itself becomes another form of editing individual shots together.

20. Susan Jeffords, *Hard Bodies: Hollywood Masculinity in the Reagan Era* (New Brunswick, N.J.: Rutgers University Press, 1994), 148.

21. Michel Foucault, *Le Souci de Soi* (Paris: Gallimard, 1984).

22. Jeffords, 157.

23. Ibid., 160.

24. Thomas B. Byers, "History Re-membered: Forrest Gump, Postfeminist Masculinity, and the Burial of the Counterculture," *Modern Fiction Studies* 42, no. 2 (1996): 431–32.

25. Ibid., 427–28.

26. bell hooks, in *Reel to Real: Race, Sex, and Class at the Movies* (New York: Routledge, 1996), points to a similar structure in the historical film of Spike Lee, *Crooklyn* (1994), although here naturalism, rather than morphing, does the trick (44–45). In *Forrest Gump,* ironically enough, the allegorical moment of reproductive morphing works counter to Masculinity. Forrest Gump speaks at the rally against the Vietnam War in Washington. One thousand extras were hired to represent the march. Then, through morphing, warping, and compositing, the mass of one thousand people was turned into two hundred thousand. Thus counterculture becomes the only allegorical moment in which the nation is morphed as such national mass body. See Parisi, 39.

27. The term "surplus value" is also meant in the Marxist sense, since technology is used to exploit labor power for the profit of a specific group: people privileged by the hegemony of Masculinity.

28. Vivian Sobchack, "Child/Alien/Father: Patriarchal Crisis and Generic Exchange," in *Close Encounters: Film, Feminism, and Science Fiction,* ed. Constance Penley et. al. (Minneapolis: University of Minnesota Press, 1991), 3–31; Jeffords, 156–77.

29. Claudia Springer, in *Electronic Eros: Bodies and Desire in the Postindustrial Age* (Austin: University of Texas Press, 1996), notes that the iconography of both Terminators is oppositional: Schwarzenegger's masculine and hard, Patrick's feminine and fluid (112). I believe that her analysis, rather than contradicting, enhances my present and later psychoanalytic reading: each Terminator encompasses both Masculine and feminine traits.

30. I owe my following remarks about ethnicity in *The Mask* to Susan Green.

31. Martin Walker makes the same point in a snappier and more mordant way in "Making Saccharine Taste Sour," *Sight and Sound* 4, no. 10 (1994): "It [*Forrest Gump*] is a sum of newsreels rewound and made blandly, briskly suitable for the age of 'Headline News,' seconds of archive footage here and brief soundbites there" (17).

32. Laura Mulvey, "Visual Pleasure and Narrative Cinema," in *Feminism and Film Theory,* ed. Constance Penley (New York: Routledge, 1988), 57–68.

33. Steve Neale, "Prologue: Masculinity as Spectacle: Reflections on Men and Mainstream Cinema," in *Screening the Male: Exploring Masculinities in Hollywood Cinema,* ed. Steven Cohan and Ina Rae Hark (New York: Routledge, 1993), 16.

34. However, the camera also tracks and zooms to compensate for the often static effect created by the long shot and the shot-in-depth. In a way, the camera adopts a changing and mobile *historical* point of view, rather than Forrest Gump's, to *compensate* for the immobility and lack of agency generated by (and as) Forrest Gump's masochism.

35. Gilles Deleuze, "Coldness and Cruelty," in *Masochism,* trans. Jean McNeil (New York: Zone Books, 1991), 7–138.

36. The only president not shown is Jimmy Carter. Apparently the lack of violence during his presidency would explain his absence from the film.

37. Byers, 432.

38. Magid, 45.

39. Fyodor Dostoyevsky, *The Idiot,* trans. David Magarshack (Baltimore, Md.: Penguin Books, 1955).

40. It is hardly coincidental that angels, alongside aliens, have made a comeback in films of the late 1980s and 1990s. Besides *Date with an Angel* (1987), films such as Travolta's *Michael* (1996) and *Phenomenon* (1996), or *The Preacher's Wife* (1996) with Denzel Washington, are the best examples of this angelical take on Masculinity. *Ghost* (1990) is also part of the genealogy of angelical Masculinity.

41. Peter N. Chumo II, "'You've Got to Put the Past behind You Before You Can Move On': Forrest Gump and National Reconciliation," *Journal of Popular Film and Television* 23, no. 1 (1995): 2–7. For a discussion of the genealogy of sanctity and subjectivity from the 1950s to the 1980s, see Vivian Sobchack, *Screening*

Space: The American Science Fiction Film, 2d ed. (New Brunswick, N.J.: Rutgers University Press, 1997), 288–89. See also Victoria Johnson, "The Politics of Morphing: Michael Jackson as Science Fiction Border Text," *Velvet Light Trap* 32 (1993): 59.

42. The success of films such as *Dumb and Dumber* (1994) is part of this agency-less idiotic masculinity. Tom Hanks's previous portrayal of a child-man in *Big* (1988) also prefaces his selection for the role of Forrest Gump.

43. Kaja Silverman, "Fragments of a Fashionable Discourse," in *Studies in Entertainment: Critical Approaches to Mass Culture,* ed. Tania Modleski (Bloomington: Indiana University Press, 1986), 137–52.

44. This is probably the main difference between *Forrest Gump* and another historical film, *Born on the Fourth of July* (1989). Given the political and dystopian effect that the latter film wants to create, renunciation of reality is not necessary. Masculinity does not need to morph with history. Ron Kovic (Tom Cruise) appears on TV only once, but in cross-cutting with TV images of Nixon. The technological difference between the new and old footage is evident to the eye, and thus any possibility of morphing is canceled.

45. One could also read *Forrest Gump* as Masculinity's turn to "technological drag." That is, one could easily read Masculinity as a subset, a technological (queer) morphed case, of drag. Drag then would be the starting point to rethink Masculinity. From here, it is only one step further to seeing *The Terminator* as a drag queen. For the way morphing has had similar effects in the case of another icon of semimasochist and posthistorical masculinity, Michael Jackson, see Johnson. She wonders: "In what ways does his tech-noir morphing inability with its 'deep emotional investment in the myth of sameness' inscribe 'cultural plasticity' as *apolitical* representation?" (63). For the relation of morphing and hyperreality to literature, see Stephanie A. Smith's discussion of Octavia Butler's *Xenogenesis* series: "Morphing, Materialism, and the Marketing of *Xenogenesis,*" *Genders* 18 (1993): 67–86. Butler's materialist embrace of the possibilities of genetic morphing points to an interesting direction in which the politics of morphing could be taken.

46. I believe that the area of domestic morphing (for instance, morphing made on PCs by private individuals using their own material, such as personal pictures) already points to a possibility to politicize morphing. When this morphed personal material is posted in Web pages and circulated through the Web, all sorts of interesting issues arise (copyright, commodification and morphing of the self, etc.). The most obvious and easiest case to study is what one could call "the morphed autobiography." On this issue, see Karen Freifeld for a comprehensive example:

"Say 'Cheese' . . . and Click Your Mouse: How to Turn Your Computer into a Digital Darkroom," *Newsday,* 24 March 1996, Nassau and Suffolk ed., A51.

47. Vivian Sobchack, "Introduction: History Happens," in *The Persistence of History: Cinema, Television, and the Modern Event,* ed. Vivian Sobchack (New York: Routledge, 1996), 3.

48. Donna J. Haraway, "A Cyborg Manifesto: Science, Technology, and Socialist-Feminism in the Late Twentieth Century," in *Simians, Cyborgs, and Women: The Reinvention of Nature* (New York: Routledge, 1991), 149–81.

Figure 10.1 "Virtual Jennifer": computergraphic makeover. Digital image by Louise Krasniewicz.

10

Orlan
Beyond the Body and the Material Morph

VICTORIA DUCKETT

The French performance artist Orlan and her performance *The Reincarnation of Saint Orlan, or Images–New Images* give fresh impetus to the question of the "morph." Involving a series of operations begun in 1990 in which she surgically remodels her face after a computer-synthesized "ideal" based on features taken from women in famous artworks (Botticelli's Venus, da Vinci's Mona Lisa, Boucher's Europa, Gerôme's Psyche, and a School of Fontainebleau Diana), this performance graphically details the conflicts unleashed and the issues raised by the notion of "morphing" as both an inscription and transformation of bodies—women's bodies in particular. It is these acts of inscription and transformation, rather than the fashionable hermeneutic through which they might be discussed, which is of compelling importance to the contemporary viewer. As Willibald Sauerländer, in his article "From Stilus to Style: Reflections on the Fate of a Notion" reminds us, "Style . . . detaches from . . . statues and images what may have been their original message and function and above all their inherent conflicts, the stamp of superstition and cruelty, the token of suffering or the signs of revolt, reducing them to patterns, samples, to the aesthetic irreality of the labeled mirror image."[1]

The most crucial aspect of Orlan's performance is a foregrounding of the "cruelties" and "revolts" otherwise elided by the smooth seduction and "aesthetic irreality" of the computer-generated morph. Hence we learn that the unveiling of Orlan's "new" face, completed after the seventh operation (in the performance *Omnipresence*, New York, 21 November 1993), was canceled because "problems with the first operation had necessitated another," and that "she fielded audience questions, her speech slowed by stitches."[2] Embodied revolts against transformation are, of course, hardly unique to Orlan, and in this essay I should like to cast her work as both material counterpoint to the digital morph and critical counterpoint to the cosmetic morph. Indeed, the disjunction between her "operationalized" material performance and the imagination of the computer-realized cosmetic morph is highlighted through this very issue of contingent bodily response. Whereas the unpredictability of Orlan's bodily responses to her operations are integrated into her performance, these same contingent bodily responses are a major subject of concern for the American Society of Plastic and Reconstructive Surgeons. Hence the emergence of a disclaimer—the "Electronic Imaging Disclaimer"—provided for physicians who use computerized images in preoperative consultations; the disclaimer requires the client to read and sign the following statement: "I understand that because of the significant difficulties in how living tissue heals, there may be no relationship between the electronic images and my final surgical result."[3] What this disclaimer marks is an awareness of the seductive qualities of digital transformation and, more importantly, an attempt to litigate—an attempt to control instigated precisely because we cannot control—the material recalcitrance of the lived body.

This very distinctiveness and recalcitrance of the material body is brought to the fore in Orlan's performances. Rather than present her transformation as a fluid, seamless, and scarless transition between two disparate images (a transition popularized in the press through their reliance on sequential "before" and "after" shots), Orlan inserts the camera into the operating "theater" to document precisely those signs of suffering and conflict otherwise absented by the (cosmetic) morph's "special effect." Forced to witness Orlan's surgery while she is conscious, while she is reading from texts and directing the action, but otherwise appearing as what she calls a "cadaver under autopsy,"[4] the audience is given access to something that is usually hidden and, indeed, doubly celebrated for its invisibility—that is, both for its invisibility of process and its invisibility of result. Here, instead, the labor and cruelty are made explicit—they are in your (her) face. As one viewer, watching the seventh operation in New York, described: "The surgical moment arrived: Orlan, lying down, is injected by a long needle under her scalp. (Camera zooms in.) But this is no simulacrum of an operation, it's the real thing. Soon, the surgeon is sawing away, methodically scraping out flesh from below the hairline. The gallery empties of a third of its audience."[5]

The difficulty experienced in watching such a performance translates into a difficulty not only in watching a mortification of real flesh but also in watching labor articulated—difficulty in watching the construction of the product rather than its fetishistic display. As Laura Mulvey tells us: "The fetish necessarily wants history to be overlooked. That is its function."[6] Orlan's surveillance camera works against this fetishistic oversight, forcing a return to temporal reference and an acknowledgment of the marks of labor. This foregrounds the "embodied" history otherwise obfuscated by cosmetic surgery and the digital morph that encourages it. A television news "special interest" spot on KACTV (Los Angeles) in late February 1996 illustrated this computergraphic encouragement of cosmetic surgery by presenting a "thinning specialist"—one Randy Rose—whose job it was to digitally "downsize" computer images of overweight people. Working for a company called Slim Photo, Ms. Rose "chipped away those unwanted pounds" in Photoshop, making her subjects "thinner without simply shrinking the body." Using a photograph of a Slim Photo client—Susan Chase—and showing the results of the computergraphic transition between an overweight "before" and a slim "after" image, the difficult and laborious physical process of transformation is elided. Although the service emphasizes that it is designed to function merely as "motivator" to bodily change, the process of this change and what it might actually entail is itself absented. Hence Ms. Chase smiles and—obviously pleased with her "new" image—says, "Wow, who's that fox? . . . So this is what I really look like under all this!" What she fails to recognize is that the process of "chipping" away those unwanted pounds can also be literalized; that the transition so smoothly and effortlessly presented on the computer monitor can itself be applied to and experienced by the material body, albeit hardly so smoothly and effortlessly. Indeed, it is precisely the literalization of this process that is being promoted—even if not explicitly—by Ms. Rose and the digital Slim Photo process.

The preoperative desire to obfuscate the fact of labor extends into the later desire to obfuscate even a trace of labor once cosmetic surgery has been completed. Thus a chart in the January–February 1996 issue of *Mirabella* listed, among other information about such surgery, the expected length of absence from work—that is, the expected time it would take for the body to cover over marks of surgical intervention.[7] In a similar vein, an article that appeared in a January 1997 issue of *Star* asked, "How do celebrities manage to have plastic surgery yet never get caught with black eyes and bandages?" and went on to explain how "the aptly named Hidden Garden in Beverly Hills is a favorite with plastic-surgery patients. It sneaks celebrities into a plush Bentley or Rolls-Royce that delivers them—via a private garage—to a door that leads directly to their room where they hide out in pampered luxury until the swelling subsides."[8] This notion of the "hidden" body is accentuated through the photographs that accompany the text: the camera that has entered this forbidden "garden" of recovery has caught only images of a veiled and anonymous female figure. Accordingly, only the recovered and smiling faces of a postoperative Ivana Trump and Jacqueline Onassis are shown and identified. This elision of the fact of physical labor and pain is again noted, although in a different context, in a related article entitled "Knifestyles of the Rich and Famous": "When actors receive awards and thank everyone from their kindergarten acting teacher to their latest agent, there's one person they are usually leaving out—their plastic surgeon."[9]

Perhaps the most interesting aspect of this absence is the fashion in which it becomes inscribed as a performance, as something that is enacted outside an individual's "lived" reality. As Nola Rocco, the owner of Hidden Garden, explains: "If you want to come as Dick Tracy or Marilyn Monroe, that's fine. Sometimes we have four Dick Tracys and four Marilyn Monroes."[10] This use of an assumed name as shield to a lived identity, this morph into some desired other, reinscribes the notion of plastic surgery as performance. In contrast to Orlan, however, this "performance" is shielded from the public gaze, and there is no audience. Hidden, the surgery is absented into a name (Marilyn Monroe) that is, interestingly, itself a "screen" to someone else (Norma Jean Mortenson). This use of an assumed name as shield to a lived identity was, of course, given antecedent through soap-opera actress Jeanne Cooper's actual on-the-air facelift in her role as Katherine Chancellor on *The Young and the Restless* ten years ago.[11] The body bearing signs of surgical intervention is thereby presented to the public gaze only when it is incorporated into the story line as a necessary fiction. Those in the real world who have cosmetic surgery usually choose to hide the surgical intervention and recuperate hidden from public view. In either case, this "absence" becomes, however, a "presence" of sorts through the attention paid to the person's absence from work, or to the "before" and "after" shots of celebrities displayed in the popular press. The "task" for the watching public thus becomes an almost gamelike matter of "filling in the blanks." Hence, beside forty-nine shots of celebrities, the *Star* explains that "sex goddess Raquel Welch would like us to think that exercise alone has kept her flesh firm, but cosmetic surgeons say she's had a facelift, a brow lift, an eyelid-lift, and some nasal resculpturing";[12] "Just before Donald dumped Ivana for Marla, the first Mrs. Trump erased 10 years with a facelift, a nose reshape, and having her wrinkles removed. Ivana attributed the transformation to a new hairstyle and lighter make-up";[13] and "Former Playmate of the Year Anna Nicole Smith says her breasts ballooned when she was pregnant with her son. Almost everyone else believes her big bosom is the result of implants."[14]

In its spectacularization of the operating "theater"—the very foregrounding of the choreography applied to each operation—Orlan's performance stands in contradistinction to the implied "magic" of cosmetic surgery. Correspondingly, one sees a macabre reversal of the cosmetic morph's status and power as special effect. In this way, Orlan has at once refused and conceded to her own moment of illusionistic grandeur: on the one hand, there is no fetishistic ellipsis of her suffering and labor, yet on the other there is a certain accommodation to the morph's theatrical construction as special performance, to its status as visual and marked effect to be watched and marveled at in its moment of unfolding and revelation. This attention paid to performance works literally to inscribe the operating room as theater, as a space in which to watch events unfold as spectacle. Orlan's *Operation One* included, for example, an African male striptease dancer and gowns designed by Paco Rabanne, and *Operation Seven* was beamed live around the world via satellite and featured Orlan interacting on the telephone or responding to faxes sent to her from places as far afield as Montreal, Tokyo, and Latvia ("Does it hurt?" asked someone from Moscow).

Although these operations were indeed performative, there must, however, also be an acknowledgment of the centrality ascribed by them to narrative; that is, the emphasis Orlan gives the "theatrical story." It is in this realm that the distinctions between the digital and material morph are best detailed: whereas the computergraphic morph represents a pause in narrative continuity or constitutes in itself a "distracting" micro-narrative and is tied, inevitably, to a reemergent cinema of attractions,[15] Orlan's "biological morph" represents a certain narrative continuity whereby each of her operations becomes a climax to a story that is in the process of unfolding. In this way, her operations can be regarded both as a concession to the "performative" nature of the morph and also as a moment in a narrative trajectory. Accordingly, the moment of transformation is staged—it foregrounds its own "special effects"—while Orlan's character is developed narratively through her interactions with a watching audience and through the texts she chooses to read to them. Hence, just as Tom Gunning explains that the cinema of attractions "is an exhibitionist cinema . . . a cinema that displays its visibility, willing to rupture a self-enclosed fictional world for a chance to solicit the attention of the spectator,"[16] so too might it be argued that Orlan ruptures the fiction of this visibility, foregrounding narrative as a tool with which to critique today's "Spielberg-Lucas-Coppola cinema of effects."[17]

Given that Orlan must be administered a dangerous spinal injection—an epidural block—to retain consciousness during the operations, these narrative insertions involve a very "real" life-and-death scenario. In this way, they stand as narrative "climaxes" within the unfolding narrative that is Orlan. They also point, somewhat more metonymically, to the questions that plague Orlan's narrative teleology: when all the operations are completed, and the time comes for her to reapply for citizenship, will the Public Prosecutor accept her new identity with her new face?[18] Will Orlan, in effect (in special effect), be allowed this complete change of identity, or will it be denied, so that she remains forever suspended in a stalled process of narrative denouement? This issue of "public" acceptance is significant because it again articulates a fact often elided: it takes public acceptance—and, more generally, public recognition—for the material morph to "fix itself."

This process of explicitly narrativizing what is generally mystified as hidden spectacle (cosmetic surgery) or as special effect (computergraphic morph) is further emphasized through Orlan's use of kitschy, Hollywood-style billboards in which she appears, divalike, surrounded by cast and production credits. Publicly documenting her process of "becoming," these stand as episodic vignettes, as literal signposts of the story's unfolding. They also join other images of Orlan. As Barbara Rose comments: "Her features and limbs are endlessly photographed; in France she appears in mass-media magazines and on television talk shows. Each time she is seen she looks different."[19] In this regard, correlations might be drawn between Orlan and her critique of the special effect and Laura Mulvey and her critique of the fetishization of the female in Hollywood cinema. Indeed, Mulvey's recognition of the way in which "the presence of woman is an indispensable element of spectacle in normal narrative film, yet her visual presence tends to work against the development of story-line, to freeze the flow of action in moments of erotic contemplation" has been both conceded and challenged by Orlan's development of narrative.[20] In the same way, Mulvey's statement that "psychoanalytic theory provided the investigative gaze (in the early 1970s) with the ability to see through the surface of cultural phenomena as though with intellectual X-Ray eyes" has met its contemporary equivalent.[21] The difference, of course, is that for Orlan, "investigation" is "intervention," which translates from the French, quite literally, into "surgical operation." Hence, Mulvey's fetishistic scopophilia, inserted as a "pause" within the sadistic voyeurism of Hollywood narrative (i.e., a close-up of the face of Garbo), is transcribed into Orlan's billboards, which stand, in their turn, as contemplative tableaux amid a like exercise in narrative demystification. Related to this *mise en abyme* of contradiction is the fact that "Orlan" has been deliberately evasive about her past, refusing specific autobiographical detail in order to maintain the mystifying "star quality" necessary for her narrative unfolding.[22]

This inscription of the biological morph as a vehicle for narrative denouement is reiterated through Orlan's use of the Renaissance painting as model to her morph. Orlan asks that the audience "not be fooled by the images but to keep on thinking about what is behind them." She tells them:

> I am sorry to make you suffer, but remember, I am not suffering, except like you, when I look at the images. Only a few kinds of images force you to shut your eyes: death, suffering, the opening of the body, some aspects of pornography for some people, and for others, giving birth.... In showing you these images, I propose an exercise which you probably enact when you watch the news on TV: do not be fooled by the images but to keep on thinking about what is behind them.[23]

This demand—that the audience "read" the image—again emphasizes Orlan's focus on "story." Relevant here is that the paintings on which her operations are based were chosen for their narrative content and not for their "inherent" aesthetic "appeal." As Orlan explains:

> I chose [the goddesses from Greek mythology] not for the canons of beauty they are supposed to represent (seen from afar), but rather on account of the stories associated with them. Diana was chosen because she refuses to submit to the gods or to men, she is active and even aggressive, she directs a group; Mona Lisa was chosen as a beacon figure in

> the history of art, a key reference, not because she is beautiful according to contemporary criteria of beauty, since beneath this woman there is a man, who we know is Leonardo da Vinci, a self-portrait hiding in the image of the "Mona Lisa" (which brings us back to the question of identity). . . . These representations of female figures act as part of my inspiration and underlie it in a symbolic manner. Their images, in relation to their stories, may re-emerge in later works .[24]

Within such a context, Orlan's face becomes representative of a "materialized" Web page in which each section asks that you point to it, click on it, and follow it through its archaeology of references. The reliance on language within this "archaeology" is further accentuated by the fact that Orlan quite literally frames the flesh removed during her operations in language, in shatterproof glass sheets on which are engraved, in various languages, an extract by Michel Serres:

> What can he do to us now, the tattooed running monster, an ambidextrous, hermaphrodite, mixed race monster? Yes, blood and flesh. Science speaks of organs, of functions, of cells and molecules, finally admitting that it is high time that we stop speaking about life in laboratories, but it never speaks about the flesh, which only just indicates the mixed, in a given part of the body, here and now of muscles and blood, of skin and hair, of bones, of nerves, and of diverse functions, which therefore mixes together that which relevant knowledge analyzes.[25]

Orlan's flesh will continue to be contained within and by this "language" until the exhaustion of her body, until she has no more flesh to be "preserved."

This attention paid to the preservation of "the fragment" and its containment in language transcribes, again quite literally, the poetic glorification of the anatomical fragment found in that Renaissance form of poetry, the *blason anatomique.*[26] Such an analogy is itself alluded to in Orlan's work through the fact that the surgery that she has been undergoing since 1990 is based on sections of Renaissance paintings. Hence she will acquire the forehead of the Mona Lisa, the eyes of Diana, the lips of Europa, and the chin of Venus. The two (the paintings and the poems) were themselves connected to a broader "preoccupation with the structure of the human body, which led to the formulation of the theory of proportions and the articulation of the scientific system of anatomy." Such a system, based on the part as the measure for the whole and on the whole as a measure of the correspondences between these parts, grounds Orlan's work within a historical continuum that she herself locates, recalling that the ancient Greek artist Zuexis chose the "best" parts from different female models and combined them to produce the ideal woman.[27] This emphasis on the correspondence of parts is, we must note, still an impetus to plastic surgery today: as a plastic surgeon advises his colleagues in an article entitled "Art for Head and Neck Surgeons," familiarity with classical art theory will allow them better to "judge human form in three dimensions, evaluate all aspects of the deformity, visualize the finished product, and plan the approach that will produce the optimal result."[28] In such a context, it is hardly surprising that this art historical and sculptural "ideal" also emerged on the pages of the *Star* with explicit reference made, for example, to "Chiseled Stone."[29] Under a banner asking "Is This the Perfect Woman?" a two-page spread of a computergraphic morph was presented and explained thus:

> Many women long for Jane Seymour's to-die-for chestnut tresses. Liz Taylor's violet eyes hypnotize. Bo Derek's chiseled cheekbones rate a 10. Those in the know admire Sharon Stone's delicate nose. Kim Basinger's pouty lips are poetry. Uma Thurman's dainty chin is in. Claudia Schiffer's great gams make men go weak in the knees. And Cindy Crawford's breasts are best. Put them all together and what do you get? With a little computer magic, we combined these highly prized, celebrity features and came up with this amazingly striking creature.[30]

This creation of the composite image also finds antecedent in the practices employed in the making of scientific images at the turn of the century. As Lorraine Daston and Peter Galison explain in their article "The Image of Objectivity," Johannes Sabotta, one of the great German anatomists of the period, amalgamated fractional parts of different individuals as the basis from which drawings would be made in his 1907 *Atlas and Textbook of Human Anatomy*. Although Sabotta's attention was devoted to the question of scientific reproduction and the "means of squelching the subjectivity of (the draughtsman's) interpretation," it is interesting to note the fashion in which this mosaic of body parts was presented as a clear image, as a tool for the comprehension of data.[31] Unlike the computer-generated morph, which positions the mosaic of parts as an uninterrupted and opaque interlude between two disparate images, Sabbota's practice might be paralleled to the act of arresting the morph in its moment of denouement. In this regard, Orlan herself can be seen as a hiatus of sorts: she is devolving into her computer-generated composite model and, as such, represents the arrested "moment" in which one image devolves into fragments of the next. Hence she graphically details the computer-generated morph's unfolding. Perhaps more important, however, Orlan also foregrounds the way in which mechanical precision—here cast in terms of computer software, but for Sabotta represented by the photograph—is less a movement toward the separation of man and machine than it is a movement toward the embroilment of one into the other. Accordingly, Orlan refers to herself as a "replicant" and regards her body—her work—as "it."[32]

Somewhat paradoxically, Orlan's very process of "objectifying" her lived body ("it") exemplifies the critical issues involved in fashioning the body after a computer-generated composite. Indeed, Orlan at once details and refuses the remodeling of the female body according to the mathematization of form. Relevant here is Francette Pacteau's discussion of the way in which "behind the woman there is, always, the image to which the question of her beauty must be referred. As beautiful as. . . . Hence desire reaches beyond mere flesh to painted perfection, which, impossibly, it wills to become flesh."[33] Orlan uses the Renaissance painting and the computer-graphic morph as generative model to her surgical operations and is therefore complicit in this process of mathematizing the female form. However, Orlan's refusal to order her face along a grid of measurements—her very asymmetry—serves to critique this technologically generated mathematization of form. Hence, two implants, usually used to emphasize cheekbones, have been placed on the temples on either side of her forehead, and in her next operation, she will be given the biggest nose technically possible in relation to her anatomy. In this way, she states, her work "is not intended to be against plastic surgery, but rather against the norms of beauty and the dictates of the dominant ideology which is becoming more and more deeply embedded in female . . . flesh."[34] Here it is interesting to note that Orlan has had to enlist the services of a "feminist surgeon" (Doctor Marjorie Cramer) for her more recent operations—the seventh, eighth, and ninth—since she "was not able to obtain from male surgeons what she was able to achieve with a female surgeon for they (the male surgeons) wanted to keep her 'cute.' "[35]

It is telling, then, that Nancy Etcoff, a psychologist from MIT, observes that a supermodel's face is "geometrically normalized" or, rather, "unusually average." As Evan Schwartz explains in a recent *Discover Magazine* article entitled "Such a Lovely Face":

> Etcoff, a psychologist from MIT who's now on staff at Massachusetts General Hospital, has long been interested in notions of facial beauty. These days she regularly visits Sandy Pentland's shop at the MIT Media Lab to study the "averaged" faces that his software yields. She has found that they bear a striking resemblance to those of supermodels such as Kate Moss, one of the most celebrated faces of the 1990s. Moss, in Etcoff's opinion, looks like an androgynous 18 year-old with few distinguishing features.[36]

This mapping of a topography that disrupts the notion of the "beautiful whole" does, of course, present Orlan as counterpoint not only to Michael Jackson but also to Cindy Jackson, a woman who has been described as "the ultimate Barbie performance artist" and who has had more than twenty operations and spent $50,000 to turn herself into a living doll.[37] The attention Cindy pays to the "metrical mean" is illustrated in the following report by M. G. Lord of a meeting with her: "She wanted a fat transplant in her cheeks. 'I have dents here which need filling in,' she explained. 'See my cheek here is flat—but I have dents underneath.' 'But I can't see them,' I said. 'But they're there,' she assured [me].'"[38] This attention paid to the detail—the detail that cannot be "seen" but might be computed—and its relation to the "whole" once more reiterates the currency that the Renaissance system of anatomy has in the contemporary period. One Dr. Bookstein, quoted in an article entitled "The Statistics of Shape: A Mathematician Uses Morphometrics to Analyze the Brains of Schizophrenics," notes that morphometrics "began with the Renaissance painter Dürer . . . [who] put grids on faces and then distorted the grids and the lines drawn within them. He used this method to explore what happened to faces as the proportions of various features changed—where an ear belongs on a long face, for instance."[39] Orlan's insistent asymmetry suggests, interestingly, that she will forever remain in a skewed grid, in an unfixed moment of "becoming," suspended in an asymmetrical tribute to the morph's unfolding.

Related to this denial of stasis is Orlan's use of the fixed photograph as document to her "recovery" from her seventh operation. She presented forty-one consecutive diptychs corresponding to her forty days of "healing" and added a final "concluding" image. At the bottom of each diptych was an image of her face digitally morphed with the portraits of her art historical reference figures. Each day, the image of the day was placed alongside these reference figures with magnets. Like Marey's nineteenth-century chronophotography, these are (surgical) still frames that show the differential between successive images and reveal a "mechanism of movement": "First of all, a face with bandages, then one with colors, from blue to yellow through red. . . . On the last day the installation is complete."[40] Orlan's specification—and literalization—of her skin color, her shape, and her form stands, of course, in contradistinction to Marey's reduction of the human form to a gait mapped against a black background. What is interesting here is that Marey "was interested in establishing a record of the norm" and erased "any distinguishing details of the man performing for him," whereas Orlan's still frames chart a movement that marks her specificity and her deviance from the "norm."[41] Not only is her movement generally kept hidden from the social sphere, but each successive image reinforces the fact that it is a human figure who is propelling "the action."

This representation of duration through a series of static frames is, as Lisa Cartwright reminds us, "about physical transformation."[42] Contrasting Marey's study of human physiology to photographer Mathew Brady's compendium of body types designated as likely to be inclined toward criminal behavior, she states:

> Between Brady and Marey we see a shift from the observed and analogically classified body to the experimented upon and the digitally ordered body. With the transition from the analogic to the digital and from observation to experimentation, we also see a shift in modes of social regulation. The body once rendered innately deviant is now open to "corrective" physiological regulation and transformation.[43]

Evidently, Orlan's surgery—and cosmetic surgery more generally—is implicated in this correction of "deviance." Perhaps more interesting, however, are the very correlations that might be drawn between the physiological "cinema" at the turn of the century and the computer morph as it functions today: the two deal with bodily transformation, ask that lines be drawn on the human form to chart this transformation, and exchange depth and form for duration and process. Popular cinema at the turn of the century might also be integrated into a discussion of the computer-generated morph today. Relevant here is Cartwright's comment that "the success of the popular cinema would depend, in part, on the spectatorial reason of viewing a continuous moving image and the masking of the technology that produced this illusion."[44] What is being noted here is a coincidence of spectatorial effect whereby the "continuous moving image" (the popular cinema) today finds its equivalence in the instantaneous simultaneity of the special effect. In other words, through the computer-generated morph, we are again witness to what Scott Bukatman calls "the boundless and infinite stuff of sublime experience . . . a transcendence of . . . human limits."[45]

We have, through this transcendence, once more returned to Sauerländer's "aesthetic irreality of the labeled mirror image." This irreality explains, in part, the name "Elastic Reality" given to the popular morphing software. Sliding easily between two images, this reality is indeed elastic, reversible. It is also a reality that hides its computations in a smooth mathematization of form. In contradistinction to this, the material biological morph produces forms of mathematization that are not, of course, reversible. The crucial difference between the computergraphic cosmetic morph and Orlan's fleshy morph lies precisely here—in the distinctions that can be drawn between forms of mathematization and forms of the body. Whereas the computergraphic cosmetic morph hides the process that puts the Renaissance grid into "place," Orlan draws attention to this process. She disrupts the grid's standard assignments and, in doing so, disrupts the grid's assignment of value. Through this disruption, she foregrounds the bodily processes by which we materially (trans)form. Rather than give us a smooth ellipsis, Orlan "faces us" with a body cut open, a bruised body, a body that (even when healed) refuses mathematical standardization: the forehead of the Mona Lisa, the eyes of Diana, the lips of Europa, the chin of Venus. Following her lead, we would therefore do well to remember to "not be fooled by the images but to keep on thinking about what is behind them."[46]

Notes

1. Willibald Sauerländer, "From Stilus to Style: Reflections on the Fate of a Notion," *Art History* 6, no. 3 (September 1983): 254.

2. Lovelace, "Orlan: Offensive Acts," *Performing Arts Journal* 17, no. 1 (January 1995): 15.

3. Cited in Ann Balsamo, *Technologies of the Gendered Body: Reading Cyborg Women* (Durham, N.C.: Duke University Press), 77–78.

4. In "I Do Not Want to Look Like. . . . Orlan on Becoming-Orlan," *Women's Art* 64 (May–June 1995), Orlan writes of herself as "an image of a cadaver under autopsy which keeps on speaking, as if its words were detached from its body" (8).

5. Lovelace, 13.

6. Laura Mulvey, "A Phantasmagoria of the Female Body: The Work of Cindy

Sherman," *New Left Review* 188 (July–August 1991): 150. She continues: "The fetish is also a symptom, and as such has a history which may be deciphered."

7. E. O'Brien, "How Many Eye Jobs Before You Can't Blink?" *Mirabella* 7, no. 3 (January–February 1996): 114.

8. Anna Stewart, "Plastic Surgery Secrets of the Stars: Exclusive Look inside the Luxury Hush-Hush Hideaway Where NOBODY Knows Your Name," *Star*, 14 January 1997, 30–44.

9. Ibid., 37.

10. Ibid., 32.

11. Anna Stewart, "Ever-Young Jeanne Is Restless for Another Facelift—at 67," *Star*, 14 January 1977, 34. The article, detailing this episode, follows the "Hidden Garden" article in *Star* cited in note 8.

12. Stewart, "Plastic Surgery Secrets," 38.

13. Ibid., 44.

14. Ibid., 41.

15. See Tom Gunning, "The Cinema of Attractions: Early Film, Its Spectator, and the Avant-Garde," in *Early Cinema: Space, Frame, Narrative*, ed. Thomas Elsaesser (London: BFI, 1990), 56–62.

16. Ibid., 57.

17. Ibid., 61.

18. "When the operations are finished, I will employ an advertising agency to find me a first and second name and an artist's name, then I will get a lawyer to appeal to the Public Prosecutor to accept my new identities with my new face. This is a performance inscribed with the social fabric, a performance which goes as far as the law . . . as far as a complete change of identity" (Orlan, 10).

19. Barbara Rose, "Is It Art? Orlan and the Transgressive Act," *Art in America* (February 1993): 83.

20. Laura Mulvey, *Visual and Other Pleasures* (Bloomington: Indiana University Press, 1989), 19.

21. Ibid., xiv.

22. "'Orlan' was [a name] taken on during her teens; until recently, when the *New York Times* revealed her name to be Mireille Porter, she refused to state what it was" (Lovelace, 15).

23. Orlan, 6.

24. Orlan, 8; see also Rose, who notes, in relation to this process of disassembling the female form, how "the fetishization of the body parts imposed on women by men since antiquity did not hold true for images of the masculine ideal. For male images, ancient artists might improve on nature, but the masculine ideal did not require fragmentation" (84).

25. Orlan, 10.

26. Francette F. Pacteau, *The Symptom of Beauty* (Cambridge: Harvard University Press, 1994), 60.

27. Rose, 84.

28. Cited in Balsamo, 58, 177–78 n. 8.

29. Stewart, "Plastic Surgery Secrets," 44.

30. Ibid., 46–47.

31. Lorraine Daston and Peter Galison, "The Image of Objectivity," *Representations,* no. 40 (fall 1992): 101–2.

32. Mark Dery, in "Against Nature," *21C* 4 (1995), writes: "Orlan, who refers to herself as a replicant and who declares—'I think the body is obsolete'—conceals a not-so-secret dream to be the world's first cyborg celebrity. . . . 'Orlan's body is it,' her work" (29–30).

33. Pacteau, 31.

34. Orlan, 9.

35. Ibid.

36. Evan I. Schwartz, "Such a Lovely Face," *Discover* (December 1995): 87.

37. M. G. Lord, *Forever Barbie: The Unauthorized Biography of a Real Doll* (New York: William Morrow, 1994), 244.

38. Ibid., 246.

39. David L. Wheeler, "The Statistics of Shape: A Mathematician Uses Morphometrics to Analyze the Brains of Schizophrenics," *Chronicle of Higher Education,* 8 December 1995, A10.

40. Orlan, 9.

41. Lisa Cartwright, *Screening the Body: Tracing Medicine's Visual Culture* (Minneapolis: University of Minnesota Press, 1995). Cartwright notes: "In Marey's study of human physiology . . . the specific form and appearance of the head and body, physical proportion, and physiognomic form . . . are rendered invisible. Any distinguishing details of the man performing for Marey, save his gait, are erased from the picture. We can know neither the skin color nor the shape and form of the features of the man whose movements are tracked in the image. Indeed, the color black, present in the black cloth that covers his skin, renders the body invisible rather than functioning as a visual indicator of racial identity" (36).

42. Ibid.

43. Ibid., 35–36. The work of Francis Galton can also be seen in relation to Mathew Brady. See Brady's "Composite Portraits," in *Nature* 18 (1878): 97–100. Discussed in Daston and Galison, 102–3.

44. Cartwright, 38; italics mine.

45. Scott Bukatman, "The Artificial Infinite," in *Visual Display: Culture beyond Appearances,* ed. Lynne Cooke and Peter Wollen (Seattle: Bay Press, 1995), 267.

46. Orlan, 6.

11

Morphing Taking Shape and the Performance of Self

SCOTT BUKATMAN

> Perhaps the immobility of the things that surround us is forced upon them by our conviction that they are themselves and not anything else, by the immobility of our conception of them.
>
> **Marcel Proust**

If the nineteenth century dreamed of cinema, then the twentieth has been dreaming of morphing. As with the trope of virtual reality, morphing articulates and condenses an array of philosophical positions and some specific desires and anxieties. Its place within the public and critical imagination tells us plenty about fantasies of disengagement and reengagement with historical as well as technological realities. Like virtual reality (VR), morphing enacts many of the contradictory impulses of contemporary culture. It *embodies* them—but it disembodies them even more. Virtual Reality, through its construction of computer-generated environments into which the (properly augmented) subject could enter, promised a total spatial plasticity that exaggerated the spatial reconstructions of earlier twentieth-century aesthetic forms, especially cinema. Morphing, a computer-generated transformation of a photographically based image, brought that level of imaginary mutability to the body and self. Again, like VR, morphing had its precedents: its reshaping of perception and bodily form recalls, say, surrealist collage or the atemporal unfolded perspectives of cubism.

Morphing is illusive but also deeply elusive; its amorphousness resists recuperative attempts to nail it down. It aspires to the condition of the floating signifier, but this is its fantasy, one that has to be critically and historically examined. Around virtual reality and morphing, images of reality, identity, and history are put up for grabs by a mutability so apparently radical that these categories appear to be superseded, even obliterated. Like so many tantalizing digital dreams, morphing holds out the promise of endless transformation and the opportunity to freely make, unmake, and remake oneself.

In this essay, I want to consider coexisting various aspects of time-based morphing sequences. Morphing is, first, a way of seeing over time, an exaggeration of everyday perceptions of continuity and discontinuity. The act of perception implicates the perceiving self and is fundamental to its definition. Morphing's hyperbole thus makes it into an explicit, condensed *performance of vision.* The same enhanced temporality also enacts a *performance of memory* yielding further self-(re)generation.

It is important to contrast the phenomenology of morphing, its performative elasticity and continual remaking of the self, with the ideology of the cultural narratives that contain and situate it. Actually existing morphing, as used in cinematic science fiction, horror, comedy, and TV commercials, is more a means of reification than liberation. Morphing and other "transgendered" and "transracial" possibilities of electronic culture (what I have elsewhere referred to as "terminal identity") stand, ever more evidently, as feeble attempts to do an end run around history's place in the construction of identity.[1] Michael Rogin correctly notes that "current writing on gender, race, and popular culture celebrates the subversive character of cross-dressing for allegedly destabilizing fixed identities. Such accounts need to consider history if they are to carry conviction."[2] Here, I want to examine morphing's slippery masquerade through its performance of racial repression in Michael Jackson's *Black or White* music video (1991) and the Jim Carrey film *The Mask* (1994). Rogin reminds us that "postmodern performance"—and morphing is a postmodern performance if there ever was one—"has an unacknowledged genealogy in the mobile, protean, modernizing self." Yet such "self-effacement can serve self-promotion," he continues: "American self-making brought with it a dark underside" of enslavement, exclusion, and appropriation.[3]

Morphing the Self: Time, Perception, Memory

Morphing, first of all, shares something with time-lapse photography. Images that recorded change over time precede the development of cinema—there are daguerreotypes that record the successive stages of a solar eclipse, and later the serial exposures by Marey and Muybridge. But time-lapse takes on a new power with cinema's introduction. Jean Epstein, who (along with Dziga Vertov and René Clair) was a filmmaker obsessed by the temporal reconfigurations of which the cinema proved uniquely capable, wrote that "a short documentary film which describes in a few minutes twelve months in the life of a plant from germination through its maturity and withering to the formation of the seed of a new generation" presents "the most extraordinary voyage." The passage of time is magnified (and revealed to be relative, local, as in Einstein's understanding), and natural processes are translated to the scale of human perception. "Thus, until the invention of accelerated and slow cinematic motion, it seemed impossible to see—and it was not dreamed of—a year in the life of a plant condensed in ten minutes, or thirty seconds of an athlete's activity inflated and extended to ten minutes."[4] Time-lapse photography provided a revelatory look beneath the humanly perceptible surface of natural development.

At the same time, it also allowed a comprehension of the thing seeing as well as the thing seen, because temporal continuity implies the existence of a consciousness for whom duration occurs. Epstein comes very close to overturning Henri Bergson's premature consigning of cinema to the realm of discontinuity (based, for Bergson, on cinema's division of movement into discrete still images)—thus Epstein anticipates Gilles Deleuze by decades. For Epstein, "the cinematograph seems to be a mysterious mechanism intended to assess the false accuracy of Zeno's famous argument about the arrow."[5] Within the consciousness of the spectator, discontinuity is effaced. Movement becomes *effectively* continuous, indivisible, and is furthermore an act—a fundamental act—of consciousness. "Experience," Deleuze writes, "always gives us a composite of space and duration."[6]

Duration is founded on what must be called a *persistence of memory*—"there is no perception which is not full of memories," Bergson pointed out.[7] "Our successive perceptions are never the real moments of things, as we have hitherto supposed, but are moments of our consciousness."[8] Time-lapse insists on the relation between vision and temporality and is thus a most obvious illustration of duration and the way it can produce the continuity and history central to a sense of self. A single frame cannot illuminate or inform in this regard: the sequence alone can do this.

Much the same is true of morphing. Morphing also makes explicit the temporal nature of visual perception, and with that, it implies the centrality of history to the existence of the self.[9] And because morphing is frequently linked to one's body, as the object that morphs, the self is still more closely linked to duration. Bergson's language even prefigures morphing, as he describes the true self as one "in which *succeeding each other* means *melting into one another* and forming an organic whole."[10] While I will demonstrate that morphing continually refuses history in the traditional sense, Bergsonian *memory* continues to function: "These two acts, perception and recollection, always interpenetrate each other."[11] Without recollection, the universe could only be a set of disconnected spatial forms existing in a permanent *now*. Consciousness, which perceives space and experiences duration, makes the self and the universe at once. Bergson's language again pushes into the terrain of morphing: perception and recollection "are always exchanging something of their substance as by a process of endosmosis."[12] *Endosmosis,* according to the *OED,* is "the passage of a fluid 'inwards' . . . to mix with another fluid on the inside of it" and so is not dissimilar from morphing, wherein elements of one picture mix with those of another.

The complex significance of this interpenetration is most apparent when a *crisis* of memory occurs, when memory endosmotically pushes itself out to morph the physical world into alignment with its own "substance." The most extraordinary literary example occurs at the start of Proust's *Remembrance of Things Past* as Marcel awakens: "I lost all sense of the place in which I had gone to sleep, and when I awoke in the middle of the night, not knowing where I was, I could not even be sure at first who I was." Marcel describes a separation of self from the objects of the world: "I would . . . open my eyes to stare at the shifting kaleidoscope of the darkness, to savour, in a momentary glimmer of consciousness, the sleep which lay heavy upon the furniture, the room, the whole of which I formed but an insignificant part and whose insensibility I should very soon return to share." But a more lingering insomnia gives him "only the most rudimentary sense of existence": "I was more destitute than the cave-dweller."[13]

Marcel is rescued by a powerful act of memory, but before allowing him the luxury of reorientation, memory first unsettles him further. "The memory—not yet of the place in which I was, but of various other places where I had lived and might now very possibly be—would come like a rope let down from heaven to draw me up out of the abyss of not-being." A great gathering of times and places occurs, drawn from memory, history, and even books recently read: "Everything revolved around me through the darkness: things, places, years." Temporal rhythms jumble: "*In a flash* I would traverse centuries of civilisation, and . . . would *gradually* piece together the original components of my ego."[14] Marcel's body, distracted by its own weariness, is initially an unreliable guide to memory's morphoses:

> Its memory, the composite memory of its ribs, its knees, its shoulder-blades, offered it a whole series of rooms in which it had at one time or another slept, while the unseen walls, shifting and adapting themselves to the shape of each successive room that it remembered, whirled round it in the dark. . . . Then the memory of a new position would spring up, and the wall would slide away in another direction.[15]

In this "waking dream" of "shifting and confused gusts of memory," the "various suppositions of which it was composed" have the indivisibility of lived experience.[16]

The neurologist Oliver Sacks has written the case study of a Mr. Thompson who could summon nothing from his long-term memory; his loss of *duration* manifested itself in a frantic remaking of his life from moment to moment. "He would whirl, fluently, from one guess, one hypothesis, one belief, to the next, without any appearance of uncertainty at any point."[17] Sacks points to the obvious need for a life story to bind our moments of experience and cites Luis Buñuel on memory: "Life without memory is no life at all. . . . Our memory is our coherence, our reason, our feeling, even our action. Without it, we are nothing."[18] Without memory, Mr. Thompson must keep morphing:

> Abysses of amnesia continually opened beneath him, but he would bridge them, nimbly, by fluent confabulations and fictions of all kinds. . . . Mr. Thompson, with his ceaseless, unconscious, quick-fire inventions continually improvised a world around him—an Arabian Nights world, a phantasmagoria, a dream, of ever-changing people, figures, situations—continual, kaleidoscopic mutations and transformations.[19]

Proust and Sacks both describe vertiginous self-transformation as a function of memory's temporary or permanent loss. Unstable memory transforms the perception of physical reality. Contemporary science fiction, however, describes an opposite "endosmosis," a morphing that alters physical reality, which then affects memory and thus the self. In today's cyber-world of digitally produced and stored multiple realities, the mere fact of physical existence no longer guarantees the *persistence* of a fixed self. Increasingly, the *sign* of memory replaces actual memories. In *Blade Runner* (1982), replicants are indistinguishable from humans because they have been programmed with memories and given photographs: visual and tangible totems of artificial remembrances. But of course, it is worth "remembering" that even for humans, memories are not merely objective intrusions of the past into our present: we make them—like Marcel, we select and distort; we misremember. To a certain extent, our pasts are constructions; to a certain extent, so too are our selves. Sacks wonders what can be done for his patients with severe memory dysfunction: "Can we create a time-capsule, a fiction?"[20] This is just what *Blade Runner*'s Tyrell Corporation has done for its artificial humans—created a fiction of time and history, encapsulated in and as photographs.[21]

If memories can be produced through programming, then "reality" becomes changeable, fungible—as a function of interpenetrating perceptions and recollections, reality can be morphed. Such "reality morphing" is central to the science fiction of Philip K. Dick. For example, the characters in *Ubik* (1991) are caught in an accelerating process of reality erosion; temporality reverses its valence as first the objects, then the rooms, buildings, and city blocks around them appear in earlier and earlier manifestations. Only Ubik, a product that appears packaged in historically appropriate forms (aerosol, ointment, elixir), can briefly restore the present—Ubik cures the heartbreak of reality morphing.[22]

Reality morphing can be, and often is, linked to metamorphoses of textuality. William Burroughs morphed reality by cutting up, folding in, and otherwise altering the layout of words on the page. Philip Dick's screenplay for *Ubik* was to end with the film bubbling and burning to a halt, and Thomas Pynchon's *Gravity's Rainbow* (1993) ended with the film breaking just before the bomb hit. Reality is as ephemeral as the paper or celluloid on which it is printed. In his Pirandellian comic book series *Supreme,* Alan Moore has troped the increasingly common process of superhero history revision: each revision of a superhero's universe (from *Batman* to *The Dark Knight* and so on) is experienced by the occupants of that universe as a massive reality morph. "I fought evil in Omega City until 1941," the original Supreme tells the most recent one, "which is when my whole world disappeared! I found myself alone in an infinity of blank, white space! I was in limbo. This may sound crazy, but it was just like I'd been written out."[23] Actually, a new version of Supreme had just been written *in.* The newest Supreme visits his "hometown" to experience his newly revised existence: "I felt a long, peculiar life well up around me."[24]

In all these examples, Proust, *Blade Runner, Supreme,* it is the world that morphs and not the body. Bergson's *self,* constantly "melting," morphing, *enduring* (in the richest sense of that word), is not, in these cases, experienced as a mutating, morphing *body*. The body of the experiencing subject tends to remain a fixed and unchanging point of reference. "As my body moves in space, all the other images vary, while that image, my body, remains invariable. I must, therefore, make it a center, to which I refer all the other images."[25] The fixed body becomes the basis for continuity of the self, especially when measured against a discontinuous world.

Bergson also points out that a "slight change" in the body has a disproportionate effect on everything else:

> [The problem] might be stated as follows: Here is a system of images which I term my perception of the universe, and which may be entirely altered by a very slight change in a certain privileged image—*my body*. This image occupies the center; by it all the others are conditioned; at each of its movements everything changes, as though by a turn of the kaleidoscope. Here, on the other hand, are the same images, but referred each one to itself, influencing each other no doubt, but in such a manner that the effect is always in proportion to the cause: this is what I term *the universe*. The question is: how can these two systems coexist, and why are the same images relatively invariable in the universe and infinitely variable in perception? . . . *How is it that the same images can belong at the same time to two different systems?*[26]

Bergson was considering not morphing bodies but rather the paradoxical relations between, as Deleuze puts it, "two fundamental characteristics of duration; continuity and heterogeneity" in psychological experience.[27] Morphing exploits this seeming paradox by *turning* one into the other (a literal troping): continuity *becomes* heterogeneity. Continuities of perception and body image melt and dissolve into kaleidoscopic mutabilities. So, as with time-lapse photography, a spectacle of continual transformation produces, at the same time, the continuity of a perceiving subject.

Time-lapse photography also reveals and explores the world that lies beneath the perceptions of the unaided human sensorium, whereas morphing tends toward an obfuscating manipulation of surface; in a sense, morphing is a parody of time-lapse images. I would like to argue an analogy: perhaps as time-lapse images once revealed the underlying processes of the natural order, morphing now reveals invisible processes of electronic manipulation to our increasingly informed gaze. I would *like* to argue something like that, but I cannot. For one thing, the special effects sequences of contemporary horror and science fiction may indeed be bursting with mighty morphin' bodies galore, but in most cases it is an

Other who morphs, and not (my) self. I'm fixed, I'm OK—you're not. Subjectivity is hardly reconfigured. In its most familiar examples, morphing is too seductive, too glossy, literally too empty. Through the literalism of morphing, as used in Jackson's *Black or White* or *The Mask,* difference is erased—heterogeneity becomes continuity. And even when it does produce some surplus "knowledge-effect" regarding digital image manipulation, at this point, the pervasive complexity of the society of the www.spectacle.com calls for something more than simple reflexive gestures to explain or encompass it.

Morphing Liberation: Movement of World

In an evident reference to Bergson's notes on the cinema, Proust distinguished between the smooth movement of a running horse from the isolated, "successive positions of its body as they appear upon a bioscope."[28] *Bio-Morph Encyclopedia,* a 1994 CD-ROM by Nobuhiro Shibayama, confuses this distinction by using morphing technology to reanimate and reinterpret some of Muybridge's photographic sequences. The original sequences were, of course, produced with multiple cameras triggered in sequence to follow a movement and so divided motion into a series of discrete moments: within nature's continuity, Muybridge produced—seemingly *discovered*—images of discontinuity and division. His images were published and often exhibited in a sequential array of discrete steps that seemed to demonstrate time's divisibility. It was this false demonstration of discrete temporal steps that led Bergson to refer to the "cinematographic effect" in negative terms. Following the invention of cinema, Muybridge and others reanimated the images, restoring an experience of temporal flow from disconnected parts.[29]

By morphing Muybridge, creating new space and time between the original images (connecting the dots, one could say), Shibayama has produced something different from either the stills or a cinematic reanimation. The original images have become a source for further experimentation rather than irreducible records of experience—they are somewhat devalued, their aura somewhat removed. Fabulously enough, Shibayama's morphing program can misinterpret what were once the stable elements of measurement within the image, such as background grid lines or some of the accidental irregularities of the ground or the movement—all these "stable," grounding elements now also warp and undulate. Shibayama also allows the user to morph chains of Muybridge sequences, bridging the gap between animal species and human genders. And so the sequences are continuous yet unstable: this is hardly a "natural" order of movement.

Morphing produces a more constant movement, more melting and transformative than discrete and intermittent. There is a continuity here, but also a marked, computer-programmed artificiality that preserves the discontinuity of the original images. The morph occurs as a real-time performance that doesn't need to be outputted to photographic stills, film frames, or any other "permanent" medium. Minor tweaking of the programming parameters can greatly alter the results. So the morph, as demonstrated by Shibayama, is repeatable, reversible, and variable, as well as—forgive me—polymorphously polysemic. But because of all this, it is also highly artificial, easily distinguished from the mundanity of lived space-time and bodily grounding, which is probably why it has been reserved for the fantastic, the alien, and the illusionist. Somewhat like the cinema, but also uniquely, morphing is both very like and unlike human perception. This new movement is as different from Muybridge's original profilmic event as from his series of discrete photographs or their reanimation. The text that accompanies the *Bio-Morph Encyclopedia* argues that "Shibayama's experiments . . . reinvest a Bergsonian sensuality into Muybridge's analyses—remapping them into a continuous flow, a pulsing world. The effect is completely different than cinematically animating them. The same time, but a very different space is regained. Fluid space. Erotic and intensional, not extensional."[30]

The *Bio-Morph Encyclopedia* is a beautiful illustration of Deleuze's writing about cinema and the "movement of world." In its articulation of duration, "cinema does not give us an image to which movement is added, it immediately gives us a *movement-image*. It does give us a section, but a section which is mobile."[31] Central to his understanding of movement are, among other things, musicals and the films of Jerry Lewis, and this leads to an emphasis, rare in film study, on performance and the movement of the body. Cinema constructs worlds of profound instability that constantly undergo transition and reconfiguration. "The shot, that is to say consciousness, traces a movement which means that the things between which it arises are continuously reuniting into a whole, and the whole is continuously dividing between things." Cinematic worlds, for Deleuze, are morphing worlds, in which movement itself is "decomposed and recomposed."[32]

In musicals, the motion of the dancer and the corresponding movements of camera, sets, and subsidiary players produce a participatory kinesis for the viewer, but also a thoroughgoing *sense* of movement of and through and between worlds. The musical produces "a mystery of memory, of dream and of time . . . a point of indiscernibility of the real and the imaginary."[33] What Deleuze observes in musicals is extended in his encounter with Jerry Lewis, a figure who emerges from the aesthetics of burlesque and musicals: "His smallest sketched or inhibited gestures, and the inarticulate sounds he comes out with, in turn resonate, because they set off a movement of world which goes as far as catastrophe . . . or which travels from one world to another, in a pulverizing of colors, a metamorphosis of forms and a mutation of sounds."[34] With Lewis, a new age is inaugurated that further "resonates" with morphing (and Michael Jackson). "This is no longer the age of the tool or machine, as they appear in the earlier stages, notably in the machines of Keaton. . . . This is a new age of electronics, and the remote-controlled object which substitutes optical and sound signs for sensory-motor ones."[35] Lewis's "new way of dancing" is involuntary, and this "movement of world on which the character is placed as if in orbit" produces Lewis's unique themes: "The 'proliferations' by which the burlesque character makes others swarm together, or implicates others who are absorbed; the cases of 'spontaneous generation' of faces, bodies, or crowds; the 'agglutinations' of characters who meet, join together, and separate."[36]

Elsewhere, I have argued that dance is where Lewis's grace and control emerge in fantastic balance.[37] Movements that first seem spastic become signals of emancipation; a precarious but exhilarating equilibrium is attained. Deleuze makes a similar point, although for him the brilliance of Lewis lies in the way he continues to function as a new sort of electronic automaton: "Even the way he walks seems like so many misperformed dance steps, an extended and recommenced 'degree zero,' with every possible variation, until the perfect dance is born."[38] For me (and apparently for Deleuze) Jerry Lewis anticipates the electronic technology of morphing, going so far as to allow himself to be endosmotically absorbed in everything around him. He, too, is polymorphous and polysemic—locked in a spastic struggle between the twin containments of narrative and the body and the multivalent possibilities offered by cinema and performance. The indivisible fluidity of Lewis's camera movements and his play with music and movement are always accompanied by constant divisibility: the division of self into multiples and "agglutinations" and "swarms" (combining morphing and cloning), the staccato rhythms and failed improvisations, the lack of punch lines and effective narrative resolutions. In the complexity and irresolution of Jerry Lewis, the fluid and the discrete, the indivisible and the divisible, coexist in an uncomfortable electronic "circuit": what I have referred to elsewhere as "paralysis in motion."[39] No wonder that Deleuze notes: "It is here for once, that it can be said that Bergson is outstripped."[40]

Black or White

Morphing is a mode of performance that certainly puts world into movement, but it is also a performance mode that depends on ideas of masquerade and mimicry. It is impossible to miss the connection between morphing and ethnicity in a world that has given us Michael Jackson and *The Mask.* Morphing allows a performance of ethnicity that at the same time defines and reduces ethnicity to performance.

"It's not about races / It's places . . . faces": Michael Jackson's *Black or White* remains a highly unsettling, unsettled work that morphs on every level. We first see Jackson in black and white—black hair, light skin, open and loose white shirt over white V-necked T-shirt, black belt, black pants, and black shoes with white socks (socks by Jerry Lewis?)—as he lets loose with a Tarzan yell. Surrounded by "natives," he dances exuberantly and plays a little air guitar. He and his troupe scoot into an undefined gray area where an arbitrarily placed ladder connotes "backstage-ness." Jackson dances with some Thai women, but a cut puts him amid Native Americans as suddenly we're way out west, with dancers in celebratory garb surrounded by trick-ridin', gun-totin', war-whoopin' Injun riders. A few more transitions bring Russian dancers into a sequence that resembles Jackson's own Pepsi commercial as well as Pinocchio's stage debut, in which he sang "There Are No Strings on Me" to demonstrate his near-human status.

Jackson also appears on a Vegas-y, Disney-ish set of world landmarks: Statue of Liberty, Parthenon, Big Ben, Eiffel Tower, Sphinx, Taj Mahal. Then comes the utopian climax, as a series of *gorgeous* people of different races, genders, and apparent sexual preferences morph into each other as they lip-synch, concluding with a pullback that reveals studio and camera (recalling the end of Jerry Lewis's *The Patsy* [1964]).

But it's not over yet. A black panther wanders from studio to street before morphing into Michael (wearing black jacket and a lovely rose lip gloss). He moves into a superbly edited tap solo reminiscent of Fred Astaire and Savion Glover, but the street set especially evokes Gene Kelly's double-exposure "solo" in 1944's *Cover Girl.* As in the Kelly number, significantly titled "Alter-Ego," a deserted street becomes a site for self-confrontation and an enraged vandalism that seems primarily produced by an unbearable self-loathing. Jackson dances, grabs himself, and, using a crowbar, repeatedly smashes the windshield of an abandoned car. There is no music, only the scuffing of Jackson's feet, assorted yells and screams, the occasional panther's roar, and the shattering of glass. Finally, Jackson morphs back into the black panther, who glances (glares?) at us and stalks off.[41]

In light of Jackson's subsequent legal problems, continued and increasingly scary surgical reformation, and growing reclusiveness, *Black or White*—and especially the final solo sequence—takes on added, and very real, poignance. The crotch-grabbing, glass-smashing violence so alarmed parents that Jackson apologized and removed it, leaving the it's-a-small-world, family-of-man morphing sequence as the finale. He never again attempted anything so risky, in which all the contradictions of his character were so nakedly, tormentedly displayed. As I once remarked about the many endings of Jerry Lewis's *The Nutty Professor* (1963), the multiplicity of solutions reveals, finally, that there is no solution to these issues of racial and personal identity. The utopian morphing of all the pretty people is as enchanting and irresistible as the world beat riff playing behind them, but the sunny message of hope is at least complicated by Jackson's dance. By dramatizing what seems to be a furious rejection of a benign, assigned role, by reclaiming a masculine black urban identity, Jackson unexpectedly performs a frightening *pas d'un* of isolation, entrapment, and refusal.[42]

Two passages from Deleuze, intended to discuss Vincente Minnelli musicals, have uncanny applicability to *Black or White.* "Dance," Deleuze writes, "arises directly as the dreamlike power which gives depth and life to these flat views, which makes use of a whole space in the film set and beyond, which gives a world to the image, surrounds it with an atmosphere of world."[43] The dynamism of Jackson's performance indeed gives depth and a kind of life to the Disney-fied versions of ethnic cultures on exotic display, and the video certainly uses the space of the set and the space of the world. But morphing remains metaphorically pervasive: in a kaleidoscopic set of substitutions, world becomes set, and set becomes world.[44] Meanings and identities are unfixed, but in conflicting ways. While the dance joyously surges through locations and ethnic barriers, creating a homogeneous globalism, the heterogeneity of the entire video's structure, with its nine sections, produces at the same time a more tentative, uncertain set of positions. Not even the song creates a unity—it's absent from the prologue and the extended solo dance. Not even Jackson creates a unity—he's absent from prologue, epilogue, and morphing sequence.

The initial dance provides a great, clear example of the movement of the dancer that extends to a movement of world. Jackson's presence is both disruptive (he rarely dances in unison with the others) and unifying (his music and moves bind sequence to sequence, race to race). Deleuze writes that "what counts is the way in which the dancer's individual genius, his subjectivity, moves from a personal motivity to a supra-personal element, to a movement of world that the dance will outline. This is the moment of truth where the dancer is still going, but already a sleepwalker, who will be taken over by the movement which seems to summon him."[45] Jackson's trancelike behavior in his solo section does recall the figure of the sleepwalker (consider Cesare, the pallid somnambulist in *The Cabinet of Dr. Caligari* [1919]). Yet what occurs via Jackson's dance is almost exactly the opposite of what Deleuze describes: the movement of world (through dance and morphing) *loses* its suprapersonal element and returns instead to a personal motivity. Something is being refused and regained in the final sequence, and with a deeply emotional ambivalence. It's not impossible to believe that this was what was truly disturbing about it, rather than Michael grabbing his penis.

Morphing Ethnicity: Masquerade, Minstrelsy, and *The Mask*

The morphing of Michael Jackson (in life and videos) and the morphing in *The Mask* demonstrate that morphing carries to a logical (if irrational) extreme some long-standing traditions of appropriation, masquerade, and disguise. The software program Elastic Reality has the user practice morphing between two head shots: one of a white man and the other of a black woman. And Michael Rogin has discussed the use of morphing technology on a 1993 cover of *Time* to create an ideal woman, an "all-American synthesis" that reflected the nation's racial makeup in proper proportions.[46] Even in these examples, morphing presents racial identity only to neutralize it.

The history of blackface performance reveals what fantasies of morphing are at least partly all about.[47] Robert Toll and Philip Boskin, among others, argue that elements of black culture began broadly to infiltrate white culture via the minstrel show. "Within its humorous confines," Boskin writes, "whites could peer into black culture without much anxiety, subject their stereotypes to some skepticism, cope with ambivalent racial feelings, and appreciate the nuances of the black experience."[48] Michael Rogin's view of blackface minstrelsy is rather less benign, and he convincingly links it to American frontier mythology, a mythology of domination.

While the Indian conveniently disappeared from mainstream American (especially urban) life, the presence of blacks and unassimilated immigrant groups was increasingly part of quotidian existence.[49] Black/white relations became more central but concurrently more repressed, while frontier mythology was sustained through all the Westerns that luxuriated in presenting the noble (and, incidentally, defeated and unthreatening) savage. Through gestures of inclusion (the white "man who knows Indians" or infusions of black style into white performance), blackface and frontier mythology brought together different races and ethnicities, but they exemplified and in fact created, "the distinctive feature of American multiculturalism: racial *division* and ethnic *incorporation*."[50]

As African Americans moved into northern urban areas in the later nineteenth century, they were seen as exotic but attractive, with "vibrantly expressive" mannerisms quite at odds with the prevailing social conventions imposed on "civilized" white citizens.[51] Blackface gave license to greater emotive range in white performance, but expanding white views of black culture occurred "within a preconceived formula."[52] Meanwhile white culture revealed a jealousy toward its own stereotype of the black man—his characteristic erotic "abandon" and "primitivism" were contained within a continual performance of mockery and emasculation.[53] Here, too, the furor over Jackson's onanistic crotch grabbing has its history.

The rhetoric of primitivism that informed both modern art and Harlem hot spots in the 1920s and 1930s increased black culture's visibility but kept the condescension. "Almost without exception, popular-culture writing in the 1920s treated Negro primitivism as the raw material out of which whites fashioned jazz. Savage, not polyphonic, rhythm, was heard in black music."[54] And this appropriation slash rescue extended to performance: "Burnt cork, so the minstrel claimed, gave Apollonian form to the Dionysiac African, making art from his nature."[55] There was indeed an "appropriative identification" at work; blackface became a means of cross-dressing, of participating in black "experience." But note, as Rogin does, that this is identification as containment, and it is also blackness without blacks, ethnicity without race—what James Snead has called an "exclusionary emulation."[56]

Following the nativist lead, Irish and, later, Jewish immigrant entertainers virtually took over blackface performance. By mocking the African American, as white performers had throughout the nineteenth century, more recent immigrants aligned themselves with dominant culture, outdistancing their own pariah status. Blackface furnished an opportunity to be both American and outsider, or, Rogin adds, to *become* American by *mocking* the outsider. By 1910, Jews dominated blackface entertainment, and attention was deflected from one stereotype (anti-Semitic) onto another (antiblack): in the 1920s, "the black mask of deference enforced on one pariah group covered the ambition attributed to the other."[57] Yet Irving Howe finds something more than mockery at work:

> When they took over the conventions of ethnic mimicry, the Jewish performers transformed it into something emotionally richer and more humane. Black became a mask for Jewish expressiveness, with one woe speaking through the voice of another. . . . Blacking their faces seems to have enabled the Jewish performers to reach a spontaneity and assertiveness in the declaration of their Jewish selves.[58]

Gilbert Seldes, writing on Al Jolson and Fanny Brice, found qualities of "daemonic" abandon and heat in their performances—they were *possessed:* "In addition to being more or less a Christian country, America is a Protestant community and a business organization—and none of these units is peculiarly prolific in the creation of daemonic individuals." Jolson and Brice, and we can add Eddie Cantor and the Marx Brothers, "gave something to America which America lacks and loves. . . . Possibly this accounts for their fine carelessness about our superstitions of politeness and gentility . . . [and their] contempt for artificial notions of propriety."[59]

This more generous, even heroic, view of blackface performance is easily aligned with Marjorie Garber's analysis of the self-aware flamboyance of transvestite perfomance: "Excess, that which overflows a boundary, is the space of the transvestite."[60] By "disrupting and calling attention to cultural, social, or aesthetic dissonances," transvestites mark a "category crisis" for culture.[61] The most obvious disruption is to categories of gender, but what is really at stake are all such "easy notions of binarity," creating a "crisis of category itself."[62] Michael Jackson, in his morphing between categories of male/female, black/white, child/adult, is exemplary (but, thank God, also unique).

Garber wants to emphasize "an underlying psychosocial, and not merely a local or historical effect. What might be called the 'transvestite effect.'"[63] But her intentional ahistoricism is precisely what Michael Rogin writes against: "Far from being the radical practice of marginal groups, cross-dressing defined the most popular, integrative forms of mass culture. Racial masquerade did promote identity exchange . . . but it moved settlers and ethnics into the melting pot by keeping racial groups out."[64] In the repeated scenes of performers blacking up, Rogin astutely unmasks the fetishistic face of the blackface musical: "*I know I'm not, but all the same. . . .*"[65] He echoes the case made by bell hooks regarding the suppression that ethnic citation can entail: hooks has argued that the attempt by nineteenth-century white women's rights advocates "to make synonymous their lot with that of the black slave was aimed at drawing attention away from the slave toward themselves."[66] Analogy did not produce an equivalence: as Rogin remarks of the corked-up performer, "It was whites in blackface, not mammies, who were allowed to perform the *separation* of self from role onscreen."[67]

Morphing shares a lot with blackface performance: both work through analogy. Analogy marks a similarity *(this is like that)* as it also affirms a distance *(but it is not that)*. Ortega y Gasset described analogy as a "mental activity which substitutes one thing for another from an urge not so much to get at the first as to get rid of the second. The metaphor disposes of an object by having it masquerade as something else."[68] But while blackface is a caricature, it is at least explicitly informed by specific ethnic histories and incursions. Morphing disposes of even that already repressed history.

Blackface, transvestism, and morphing all participate in analogy's subtle suppression (this is *both*, but also *neither*); but morphing brings with it a new literalism—this *becomes* that; it *takes* shape. Appropriation thus becomes an act of complete erasure. Morphing confirms the validity of Rogin's arguments in the way that it literalizes the suppression that blackface had realized with more subtlety but less thoroughness. Morphing, a celebration of endlessly transmutable surface, becomes a sign only of itself, hardly even alluding to the complexities of history and ethnic culture behind its digital gloss. By "rendering" everything as surface, and all surfaces as equal, morphing becomes *a caricature of blackface.*

The Mask provides an excellent example of Rogin's ethnicity-without-race, as ethnicity itself becomes a mask to be donned and doffed with surprising facility. In the film, shy Stanley Ipkiss, played by the extraordinarily white Jim Carrey, finds a "primitive" wooden mask, which, when worn, transforms him into a green-skinned, zoot-suited parody of African American performer Cab Calloway. The film offers an explicit and misleading interpretation of its own themes before the plot proper even begins, as the author of a book called *The Masks We Wear* is seen being interviewed on television and explaining that masking is a means of suppressing the id; in the film, however, masking functions as a means of *releasing* the id. This is only the first of many recuperations or suppressions the film uses against more liberating slippages of self.

Calloway has himself been seen as a caricature, playing the fool for white audiences. In the all-black Fox musical *Stormy Weather* (1943), for example, Calloway struts his considerable stuff while performing in an outrageously exaggerated zoot suit topped by massive shoulder pads and a bow tie at least a foot wide. The zoot suit was, by this time, fully associated with primarily Latino and black gangs in Los Angeles and other integrated urban areas. With its generous tails, wide lapels, and baggy trousers, the zoot suit was a literal gesture of defiance in times of cloth rationing, and in the Hearst press and elsewhere, zoot-suiters were portrayed as subversives at war with (invariably) white servicemen. On Calloway, however, the zoot suit became just another costume.[69] Calloway, his hair flying in all directions, is easily seen as a parody of the ethnic wild man, an unrestrained id playing—this time—for laughs. Whether this view of Calloway is legitimate—and it ignores his skills as bandleader, composer, and singer—it is true that Calloway served as quite the deracinated icon of "hep" in the 1930s and 1940s.

Calloway's own impersonation of the ethnic Other is even more interesting in light of his appearance (in both senses of the word) in three Fleischer Brothers cartoons from the 1930s. The Fleischer cartoons are significantly funkier, and vastly more surreal, than the product from other studios such as Disney and MGM: things are continually transforming, or coming to life; they exist in uneasy but wondrous states of trembling impermanence. I have to say it: they *morph.* In the Betty Boops in which Calloway performs, he appears first as "himself," and later as a rotoscoped, animated figure (rotoscoping, whereby photographic footage is traced, is an obvious kind of protomorphing). Max and Dave Fleischer were proudly Jewish New Yorkers—their cartoons are rich in ethnic flavor—and in the context of Jewish identification with black performance styles, it's not difficult to see these acts of rotoscoping in relation to blackface as yet another entry in the history of cinematic minstrelsy. (In one Betty Boop cartoon, Koko is pursued by the disembodied head of Louis Armstrong. The Hebrew word for "kosher" appears on a speedometer that emerges from Koko's *tuchus.*)

So a remarkable trajectory is described: Carrey's white Stanley Ipkiss becomes an animated parody of the self-parodic Cab Calloway, who had himself already been transformed into a cartoon version of himself in the 1930s. In *The Mask,* however, morphing works to downplay the film's evident racial subtext. Carrey's character is presented not as a jazz aficionado but as a cartoon buff, with animation cels on his bedroom wall and a statue of Tex Avery's (again) zoot-suited Wolf character all too prominently displayed. When Ipkis morphs, the film's comedy depends on a level of exaggeration usually reserved for, and associated with, cartoons: The Mask bounces off walls, floors, and ceilings; his eyes and tongue bug way out of his head; his heart pumps visibly (and is visibly heart shaped). Etcetera. In an altogether classic example of Freudian displacement and fetishism, the prominent cartoon references hide, or at least downplay, ethnic associations. Greenface trumps blackface.

The greenface/blackface masquerade of *The Mask* should also be juxtaposed with the Joker's whiteface/blackface in Tim Burton's first *Batman* film (1989). Actually, there are two blackface figures in the film: the Joker, with his pained and painted frozen Sambo smile and purple zoot suit (*again* with the zoot suit?), is once more a parody of a parody—the tragic clown, the blackface performer who can no longer remove the mask—while the more mysterious, darkly muscled figure of the Batman summons (and contains) the anxiety surrounding the Mandingo figure of black sexual power.

Based on a comic book from Dark Horse Publications, *The Mask* is also a kind of superhero remake of *The Nutty Professor*, but devoid of that film's unsettling ambiguities of identity. *The Mask* instead supplies an endless rhetoric of authenticity and sincerity: it's all about finding *the real you.* As an unchanged Ipkiss embraces an improbably attractive blonde, the finale manages simultaneously to reject ethnicity, the unconscious, morphing, *and* narrative plausibility. Unlike the more challenging and agonized works by Jerry Lewis and, indeed, Michael Jackson, *The Mask* entirely reifies the "authentic" self. In this light, Carrey's non-Jewishness is as important a suppression here as his alter ego's nonblackness: ethnicity and history are morphed out of existence on both sides. Here lie the limits of the "category crisis" that Garber emphasizes: erasure presents a category crisis, *but also its resolution.*

Performance and Possibility

The endlessly regenerative self-creation of morphing thus permits history's elision and repression. Garber's utopian vision of "category crises" must contend more fully with American mythologies of self-invention and reinvention, a history that denies history's determinism. We are not trapped by that history, but we do ourselves no service by denying it. On the other hand, Garber is hardly wrong in locating a pervasive crisis underlying Western identity formation. Historically and philosophically, resolution and closure are hardly as totalizing as they pretend to be; pluralism and diversity are never entirely subsumed.[70] If containment is the goal, that implies that some areas of excess require containing, and needless to say, attempted containment is hardly tantamount to successful containment.[71]

Garber argues that the "mechanism of substitution, which is the trigger of transvestic fetishism," is "also the very essence of theater."[72] Theatrical performance ("role playing, improvisation, costume, and disguise") is a licensed, but still unsettling, transvestism: "All of the figures onstage are impersonators."[73] Performance also extends to the acts of writing and reading. In *S/Z,* Roland Barthes muses upon the radical potentials of narrative language, a potential almost invariably denied and suppressed by particular readings of inherently pluralistic texts:

> The slash (/) confronting the S of SarraSine and the Z of Zambinella has a panic function: it is the slash of censure, the surface of the mirror, the wall of hallucination, the verge of antithesis, the abstraction of limit, the obliquity of the signifier, the index of the paradigm, hence of meaning . . . to choose, to decide on a hierarchy of codes, on a predetermination of messages, as in secondary-school explications, is *impertinent* since it overwhelms the articulation of the writing by a single voice.[74]

The text becomes a sustained performance of mis-taken identities: multiple valences and unstable analogies constantly edge up on (or transgress) the limits of assigned meaning. "Furthermore, to miss the plurality of the codes is to censor the work of the discourse: non-decidability defines a praxis, the performance of the narrator." Barthes thus argues for a utopian understanding of metaphor: "[A] successful metaphor affords, between its terms, no hierarchy and removes all hindrances from the polysemic chain."[75]

Morphing melts through the "panic function" of the slash—it is mutable, and repeatable and variable and reversible. All codes become potentially plural; they disappear at the moment they emerge. Garber writes that the transvestite performance is "always undoing itself as part of its process of self-enactment,"[76] but *S/Z* indicates that this might be true of *all* performance, albeit in less explicit ways. The duration common to semic chain, time-lapse sequence, morph, and performance provides "plurality and circularity" rather than a unidirectional and unchallenged telos: they all vibrate with possibility.

The Mask, for all its morphological play, denies this dual valence of enacting and undoing. Here, transformation reinstates the slash of censure and repression, enforcing a tyranny of decision, hierarchy, and predetermination. Compare *The Mask*'s tedious univocalism with, say, someone like Carmen Miranda (certainly a muse for transvestite culture). Shari Roberts has discussed contradictory readings of Miranda, first as an inadequate signifier of any existent ethnic identity, and then her later appropriation as a camp tutti-frutti Technicolor explosion of polysemic artificiality (*her* voice overwhelms by its hilarious incomprehensibility).[77] Miranda's playful masquerade points to the limitations of *The Mask*, where narrative containment and fixed ideas of identity, authenticity, and ethnicity work together to suppress the more polymorphous phenomenologies of morphing and masquerades. The delirious, kaleidoscopic movement of world is limited by predictability, stasis, and reification. The sense of denial that seems to be working overtime in *The Mask* is what makes it an immature, inferior work—where Jerry Lewis and Michael Jackson put more of themselves into play than they can ever possibly contain, and where Barthes and Deleuze recognize polysemic richness, *The Mask* "settles" self and world within an unchallenging and ahistorical haze. *Black or White* provides something similar, at least until Jackson smashes that neat little fantasy with a crowbar.

So the ideological determinism of *The Mask* should not, and need not, blind us to morphing's phenomenology, its performance of duration and becoming. Through its fundamentally cinematic blending of heterogeneity and continuity, morphing can become a shimmering performance of unresolved transformation. Because our existence depends on our moments of self-definition "melting into one another," transformation is fundamental to continuity. We are never fully ourselves, never fully resolved. Morphing provides a profound illustration of our own irresolution without requiring that we accept its conditional instabilities as ahistorical absolutes. Transformation is not unbounded, ahistorical, or absolute, but neither is it simply an illusion lacking in truth or reality.[78]

Conclusion

There is something symptomatic in my simultaneous desire to embrace and reject morphing. Much of my writing has been about morphing in some form or another. Jerry Lewis morphs into multiple selves, on-screen and off, in a jerky dance of simultaneous liberation and repression. Superheroes, with their secret identities, morph; and some, like Plastic Man or Mr. Fantastic, are even shape-shifters. And the terminal identities of cyber-culture promise a liberation from space, time, flesh, and history.[79]

Looking back, I can see my own desire to accumulate identities without settling down into any one—I'm always ready to morph. But in its most familiar versions, in countless sci-fi movies and TV commercials, the hollowness of morphing—its literal hollowness, the fact that there's nothing inside—offers surprisingly scant room for fantasy. Morphing is an inadequate, overly literal gesture toward change without pain, without consequence, without meaning. There is something comforting, perhaps, about the stability of unstable identity, but morphing holds out empty arms. (Or have I finally, simply, figured out that I'm stuck with myself? Maybe I'm in Supreme's position, and I need to explore the "long, peculiar life" that has somehow welled up around me . . .)

Notes

This is for Vivian Sobchack for making me do it. I would also like to thank Ira Jaffe, Diana Robin, Brian Kelly, and Deborah Jenson for their comments and ideas.

1. Scott Bukatman, *Terminal Identity* (Durham, N.C.: Duke University Press, 1993).

2. Michael Rogin, *Blackface, White Noise: Jewish Immigrants in the Hollywood Melting Pot* (Berkeley: University of California Press, 1996), 12.

3. Ibid., 51–52.

4. Jean Epstein, "Magnification and Other Writings," trans. Stuart Liebman, *October* 3 (spring 1977): 19.

5. Ibid., 17, 23.

6. Gilles Deleuze, *Bergsonism,* trans. Hugh Tomlinson and Barbara Habberjam (New York: Zone Books, 1991), 37.

7. Henri Bergson, *Matter and Memory,* trans. Nancy Margaret Paul and W. Scott Palmer (1911; reprint, New York: Zone Books, 1991), 33.

8. Ibid., 69.

9. Martin Jay, in *Downcast Eyes: The Denigration of Vision in Twentieth-Century French Thought* (Berkeley: University of California Press, 1994), writes that our selves should properly be identified with "the internal experience of individually endured time, the private reality of *durée*" (197).

10. Bergson, *Time and Free Will: An Essay on the Immediate Data of Consciousness*, trans. F. L. Pogson (1910; reprint, London: G. Allen and Unwin, 1959), 128–29.

11. Bergson, *Matter and Memory*, 67.

12. Ibid.

13. Marcel Proust, *Remembrance of Things Past*, vol. 1, trans. C. K. Scott Moncrieff and Terence Kilmartin (1922; reprint, New York: Random House, 1981), 4–5.

14. Ibid., 5-6.

15. Ibid., 6.

16. Ibid., 7.

17. Oliver Sacks, "A Matter of Identity," in *The Man Who Mistook His Wife for a Hat* (New York: Summit Books, 1985), 104.

18. Luis Buñuel, *My Last Sigh*, trans. Abigail Israel (New York: Vintage Books, 1984), 4–5.

19. Sacks, "A Matter of Identity," 104.

20. Sacks, "The Lost Mariner," in *The Man Who Mistook His Wife for a Hat*, 41.

21. For more on this film, see my *Blade Runner* (London: British Film Institute, 1997).

22. Phillip K. Dick, *Ubik* (New York: Vintage, 1991).

23. Alan Moore, *Supreme* 2, no. 41 (August 1996):13.

24. Moore, *Supreme* 2, no. 42 (September 1996):3.

25. Bergson, *Matter and Memory*, 46.

26. Ibid., 25.

27. Deleuze, *Bergsonism*, 37

28. Proust, 7.

29. For Jean-Louis Baudry, conversely, individual frames admitted a Derridean difference that projected sequences denied. See Jean-Louis Baudry, "Ideological Effects of the Basic Cinematic Apparatus," trans. Alan Williams, *Film Quarterly* 28 (winter 1974–1975): 39–47.

30. The notes are by David D'heilly.

31. Gilles Deleuze, *Cinema 1: The Movement-Image*, trans. Hugh Tomlinson and Robert Galeta (Minneapolis: University of Minnesota Press, 1989), 2.

32. Ibid., 20.

33. Gilles Deleuze, *Cinema 2: The Time-Image*, trans. Hugh Tomlinson and Robert Galeta (Minneapolis: University of Minnesota Press, 1989), 64. The films of Vincente Minnelli serve as the point of reference.

34. Deleuze, *Cinema 2*, 65. Deleuze illustrates the first example with the de-

struction of the music professor's home in *The Patsy* (1964), and the second with the transformation sequence in *The Nutty Professor* (1963).

35. Deleuze, *Cinema 2*, 65.

36. Ibid., 66. Deleuze's examples here are, respectively, the six uncles in *The Family Jewels* (1965), the three women in *Three on a Couch* (1966), and the characters in *The Big Mouth* (1967).

37. Scott Bukatman, "Paralysis in Motion: Jerry Lewis's Life as a Man," in *Comedy/Cinema/Theory*, ed. Andrew S. Horton (Berkeley: University of California Press, 1991), 188–205.

38. Deleuze, *Cinema 2*, 65.

39. Bukatman, "Paralysis in Motion."

40. Deleuze, *Cinema 2*, 66.

41. Even this is not the end: subsequently an animated framing sequence gives us Bart and Homer Simpson arguing about TV.

42. There is some parallel to the monstrous, masochistic morphosis Jackson undergoes in *Thriller* (1982). One could also see some anticipation, in its racial anger and reclamation of African American tap culture, of the polemical stage musical *Bring in 'Da Noise, Bring in 'Da Funk* (1996) by Reg E. Gaines, George Wolfe, and Savion Glover.

43. Deleuze, *Cinema 2*, 62.

44. These substitutions form another link to Minnelli's aesthetic of illusionism.

45. Deleuze, *Cinema 2*, 61.

46. Rogin, 7. The computer artist Nancy Burson had been creating similar composite images at least a decade earlier.

47. Blackface minstrel shows dominated popular culture by the turn of the century, and nearly every city and town had its troupe. Innumerable composers dipped heavily into black vernacular culture for its figurative, colloquial, and erotic languages. Blackface finally waned as a tradition in the late 1920s but remained a big part of the Hollywood film musical. Despite its seeking refuge in the cinema, the naturalistic demands of sound cinema were certainly a major factor contributing to the disappearance of this very theatrical, very conventionalized performative mode.

48. Joseph Boskin, *Sambo: The Rise and Demise of an American Jester* (New York: Oxford University Press, 1986), 93–94. See also Robert C. Toll, *Blacking Up: The Minstrel Show in Nineteenth-Century America* (New York: Oxford University Press, 1974).

49. Rogin writes: "American film was born in the industrial age out of the conjunction between southern defeat in the Civil War, black resubordination, and national integration; the rise of the multiethnic, industrial metropolis; and the

emergence of mass entertainment, expropriated from its black roots, as the locus of Americanization" (15).

50. Rogin, 26.

51. Boskin, 68.

52. Ibid., 78.

53. See Eric Lott's *Love and Theft: Blackface Minstrelsy and the American Working Class* (New York and Oxford: Oxford University Press, 1995) for more on the simultaneous attraction and anxiety revealed by white stereotyping in the nineteenth century.

54. Rogin, 113.

55. Ibid., 22.

56. James Snead, *White Screens/Black Images* (New York and London: Routledge, 1994), 60. In the introduction to her recent book *Black Talk: Words and Phrases from the Hood to the Amen Corner* (Boston: Houghton-Mifflin, 1994), Geneva Smitherman pointed out that "while Black *talk* has crossed over, Black *people* have not."

57. Rogin, 105.

58. Irving Howe, *World of Our Fathers* (New York: Simon and Schuster, 1976), 563.

59. Gilbert Seldes, *The Seven Lively Arts* (1924), quoted in Howe, 566.

60. Marjorie Garber, *Vested Interests: Cross-Dressing and Cultural Anxiety* (New York and London: Routledge, 1992), 28.

61. Ibid., 16.

62. Ibid., 10, 17.

63. Ibid., 36.

64. Rogin, 12. He also notes: "One might expect endorsements of masquerade to run aground on their largely repressed past, meeting their match in blackface. In the current excitement over popular culture, however, the direction of influence runs the other way, not disciplining present theory by past practice, but opening the past to contemporary interests" (35).

65. Ibid., 182–83.

66. bell hooks, "Racism and Feminism," in *Ain't I a Woman?* (Boston: South End Press, 1981), 141.

67. Rogin, 193.

68. Ortega y Gasset, quoted in hooks, 141.

69. Thus the *outrage* of white readers of yellow journalism directed against ethnic subcultures was transformed into the more benign and harmless posing of the comically *outrageous*. Further, because *Stormy Weather* was produced as a

government-encouraged gesture of inclusion toward black servicemen stationed overseas, the zoot suit was itself impressed, shanghaied.

70. As an alternative to Michael Rogin's rather totalizing picture of white appropriation of black culture, see Ann Douglas, *Terrible Honesty: Mongrel Manhattan in the 1920s* (New York: Farrar, Straus, Giroux, 1995). Her description of the "mongrel Manhattan" of the 1920s provides ample evidence of the ongoing interchange between black and white cultures in urban America.

71. Perhaps this is why the *performance* of cross-dressing is usually more intriguing than its literary *narration.*

72. Garber, 29.

73. Ibid., 40.

74. Roland Barthes, *S/Z,* trans. Richard Miller (New York: Hill and Wang, 1974), 107, 77.

75. Ibid., 77.

76. Garber, 149.

77. Shari Roberts, "The Lady in the Tutti-Frutti Hat: Carmen Miranda, a Spectacle of Ethnicity," *Cinema Journal* 32, no. 3 (1993): 3–23.

78. The utopianism of the musical, Richard Dyer reminds us, may be mythic, but in, say, Judy Garland's blend of theatricality and authenticity, real negotiations take place. See Dyer's "Entertainment and Utopia," *Movie* 24 (1977): 2–13, and *Heavenly Bodies: Film Stars and Society* (London: Macmillan Press, 1986).

79. See my "Paralysis in Motion: Jerry Lewis's Life as a Man," *Terminal Identity,* and "X-Bodies: The Torment of the Mutant Superhero," in *Uncontrollable Bodies: Testimonies of Identity and Culture,* ed. Rodney Sappington and Tyler Stallings (Seattle: Bay Press, 1994), 92–129.

12

Special Effects, Morphing Magic, and the 1990s Cinema of Attractions

ANGELA NDALIANIS

The event? *Terminator 2: 3D Battle across Time* (1996)—a multimedia attraction at Universal Studios, Florida. Schwarzenegger, Hamilton, Furlong, and Patrick reprise their movie roles under the direction of James Cameron. Screen action using computer, video, and film technology combines with live action within the theater to produce an exhilarating participatory entertainment experience. We (an unsuspecting group of adventurers) enter the "Cyberdyne Complex" at Universal Studios, and our tour of the installation begins. We're to be present at the unveiling of the latest Cyberdyne innovation: the T-70 (a primitive 1997 version of the T-800). While our Cyberdyne host introduces us to the company's "cybotic" vision through a video projected on multiple television screens, the video control room is invaded by characters Sarah and her son John Connor, who warn the audience (via video screens) to make a run for it. After Sarah refreshes our memories about the events that took place in *Terminator 2: Judgment Day* (1991) by narrating and showing us brief scenes from the film (and as the reality of our presence at Universal melds with the fiction of the Terminator universe), Cyberdyne once again takes control of the transmission, and an embarrassed host warns us to ignore the wild ramblings of the rebels. The audience is then ushered into a theater, and the presentation continues. As we sit dumbstruck at the sight of the "live" T-70s who flank the audience on either side of the auditorium, our state of awe continues when we're treated to the sight of Sarah and

John's further invasion (this time as "real" actors in our space) of the theater. Suddenly we find ourselves thrust into the center of a live-action battleground.

Meanwhile, behind the Cyberdyne representative onstage we see the Cyberdyne logo projected onto a twenty-three-by-fifty-foot screen. Amid the battle chaos (and with our 3-D glasses on), we become the meat in a sandwich of bullets that are fired at the Connors by the T-70s. Some of the shots fired in the auditorium hit the Cyberdyne logo on the screen. With appropriate sound effects booming out of the 159 speakers, we look on in horror as the bullet holes embedded in the logo melt and morph into the T-1000. As his liquid-blob shape morphs into "chrome man" guise, the T-1000's head begins to fill the twenty-three-foot screen, then lunges forward, seemingly escaping the confines of the screen space as it thrusts toward the audience, who parallel the motion by screaming hysterically and reaching out to protect their faces from the coming onslaught. The T-1000 then "moves back" into screen space (his 3-D form flattening back into the more two-dimensional nature of the "flat" screen format), morphing into liquid-metal form and slipping down to the lower part of the screen, where, in search of his prey, he transforms into a live actor (in the policeman form of the T-1000) onstage. Moments later, a time portal opens up in the screen, and a "live" T-800 (a Schwarzenegger look-alike) arrives on stage on a Harley-Davidson Fatboy, calling out a repeat performance of his famous one-liner to John: "Come with me if you want to live." Both then enter the screen reality (leaving Sarah behind to deal with our present reality), and the next stage of the story begins: a seven-minute film that dramatizes John and the T-800 trying to destroy the computer mainframe in the year 2029 of Cyberdyne's future.

The effects illusions in *Terminator 2: 3D* are indicative of the accelerated pace at which entertainment industries are transforming as a result of new computer technologies—altering with them the perceptual and sensory responses that these effects extract from their audience. The attraction makes the theme park experience that Spielberg's *Jurassic Park* (1993) narrativized come yet one or two steps closer to slipping into our reality. *Terminator 2: 3D*'s internal logic (like so many other examples of contemporary entertainment media) is driven by the concept of the "morph." As a new technology, morphing has become one of the major special effects figured in contemporary cinema, encapsulating many of the dramatic changes that have occurred in the film industry over the last decade. This essay is concerned with two transformations in particular.

First, the evolution of morphing parallels the morphing of the film industry itself—especially its increased adeptness and reliance on digital technology. The effects produced in films such as James Cameron's *The Abyss* (1989) and *Terminator 2: Judgment Day*, and Steven Spielberg's *Jurassic Park* (1993), put the nails in the optical-era coffin, leading to the replacement of traditional film technology such as stop motion photography, optical printing, creature effects, and matting by digital effects.[1] Furthermore, the cinema itself is morphing into other media, reflecting the impact that horizontal integration of the industry has had on 1990s entertainment. Computer games, cinema, television, and theme park attractions are now engaged in a complex level of interaction that makes it increasingly difficult to untangle one media form from another. Does *Terminator 2: 3D Battle across Time*, for example, belong to the realm of cinema, television, computer technology, theater, or theme park attraction? Where does the theatrical experience end and the film experience begin? Where does the film end and the attraction begin?

The second transformation this essay will focus on is contemporary special effects cinema's return to an aesthetic that dominated in earlier phases in film history, and in the precinema era. The technologically produced spectacle so typical of morphing—and computer technology in general—has much in common with an earlier special effects cinema that shared its capacity to produce fantastic illusions with magic traditions that were dominant in the nineteenth century.

The Morphing of a Film Industry

Like many of the effects films and attractions that preceded it, the *Terminator 2: 3D* attraction has pushed film technology and computer graphics to new limits while acknowledging its dependence on film technology of the past—in this instance, of the 1950s.[2] At one stage in the film section of *Terminator 2: 3D*, the 23-by-50-foot screen expands to 180 degrees and is flanked by two additional screens of identical size—changing the dimensions of the screen to 23-by-150 feet of enveloping spectacle. However, while harking back to the 1950s era of Cinerama, Cameron and the effects crew of Digital Domain also take this technology much further. The 70 mm projected film no longer reveals the graininess of 1950s Cinerama; film quality combines with digital effects to produce a crystal-clear depiction of an alternative reality that invades our own. Digital technology also rejuvenates 1950s 3-D cinema—as does the new screen, which is coated with high-gain material allowing for the best possible 3-D imagery.

Morphed beings are now placed within a 3-D context, and the illusionistic outcome is not only technologically groundbreaking but phenomenologically amazing. Audience members sit stunned in their seats, marveling at how these illusions are possible. Likewise, the surround sound systems that wide-screen cinema first introduced as a five-speaker format in the 1950s (and which were given new life with the 1977 release of *Star Wars* and the era of surround sound entertainment cinema that followed) are now replaced with new digital audio effects by Soundelux that comprise 45,620 watts of sound blaring through 159 speakers.[3] Furthermore, simulation ride technology works on a mammoth scale in this attraction: the norm of approximately six to twenty seats per ride found in attractions such as *Back to the Future* (Universal Studios) and *Star Tours* (Walt Disney World) is now expanded to accommodate approximately seven hundred vibrating seats. At one stage during the attraction, the auditorium floor moves, and we appear to go down an elevator into the depths of Cyberdyne with Schwarzenegger and Furlong.[4] The combined effort of all of these innovative effects makes the experience seem real.[5] That is, state-of-the-art digital effects, the digital sound system, and simulation ride technology combine with "older" technologies of wide-screen and 3-D cinema to produce an immersive and sensorially stimulating entertainment experience. By integrating the morphing effects within a complex web of cinematic, computer-generated, and theatrical effects, this attraction provides an entertainment spectacle that places the audience in the middle of the action. The perceptual barriers that separate "real" audience space from "illusionistic" cinema screen space magically appear to collapse, and (as described) the morphed image of the T-1000 works in unison with this technology to collapse the film frame that separates illusion from reality. Thus, in the words of Schwarzenegger: "The topography of motion pictures continues to change at the speed of light, becoming more and more interactive with audiences across the globe. . . . What we have created with *Terminator 2: 3D* is the quintessential sight and sound experience for the 21st century, and that's why I'm back."[6] This performance is very much a performance of state-of-the-art entertainment technology—an entertainment technology that will continue to morph itself into the twenty-first century.[7]

In this regard, it was no coincidence that genres such as science fiction, fantasy, and horror underwent a boom during the 1980s. Their revival of popularity clearly coincided with the growth in special effects houses. Given its capacity for embracing both horror and the fantastic, the genre of science fiction proved especially popular because it could exploit the illusionistic potential of special effects. Science fiction cinema—especially its early- and late-twentieth-century manifestations—has always reveled in displaying the technological capabilities of the film medium while thematically deliberating over the future effects of such technology.[8] Emerging during a period when the cinema was concerned more with the spectacle of film technology than with narrative, the beginning of the science fiction film genre has its roots in films such as Méliès's *A Trip to the Moon* (1902). This early "trick film" used its science fiction premise as an excuse to exhibit the "magic" of new film technology: the cinema's technological capabilities were presented in the form of elaborate special effects.[9] Much like the pioneering examples of the science fiction genre at the turn of the century, contemporary science fiction cinema has provided a venue for innovative developments in film technology—developments that have been depicted on-screen in fantastic terms. As examples of the genre, films such as *Star Wars, Close Encounters of the Third Kind* (1977), *The Abyss, Terminator 2, Jurassic Park,* and *The Lost World: Jurassic Park* (1997) have been at the forefront of new developments in contemporary cinema, pushing technology to new limits and showcasing it in blockbuster format—whether in the environment of the cinema or the theme park.

1990s special effects technology and audience experiences mark a radical turning point that follows a series of changes in film technology introduced in the late 1970s. The transition year was 1977, and the two films that ushered in the new era were *Close Encounters of the Third Kind* and *Star Wars.*[10] Both films introduced a new sensibility into the film experience. Traditional perceptions of film space were on their way to being dramatically altered, and since the release of these two films, particularly in the 1990s, computer-generated imagery has been integral to these new perceptions. *Star Wars* in particular breathed new life into film technology that had lain dormant since the early 1960s. Much like Cameron did in *Terminator 2: 3D,* in reviving the "entertainment-as-effects-spectacle-and-immersive-experience," George Lucas turned for inspiration to special effects and film technology that had dominated in the 1950s. For example, his effects unit revived 1950s wide-screen technology: old VistaVision cameras that had remained unused since the early 1960s, the matte painting tradition, stop motion photography, surround sound, and the art of combining models and miniatures with paintings.[11] Despite the use of primitive (by today's standards) computer technology, *Star Wars* anticipated the era of computer graphics and stands as a transition point between the old and the new.[12] Ironically, in achieving a specific kind of special effects spectacle, Lucas's revival of traditional film technology, and its melding with new technological developments devised by Industrial Light and Magic (the special effects company Lucas formed to fill a gap in special effects required by the birth of a new kind of entertainment cinema), also initiated the eventual demise of much of this traditional film technology.

In the blockbuster entertainment cinema tradition that has followed in the 1980s and 1990s, the "fantastic" has become the mode through which to explore and push to new limits the technological capacity of the cinema. With each of the innovative major effects films released in the last decade, each journey into the fantastic and science fictional has also expanded the limits of computer technology itself.[13] Each "fantastic" film effect owes its realism to new developments in computer software and hardware; and each "fantastic" special effect improves on and advances the technology that preceded it, making the fantastic appear even more realistic. Digital technology has necessitated a metamorphosis of the film industry, and morphing software (and the hardware needed to implement it) has played a key role not only in advancing computer graphics but also in advancing the illusionistic potential of the cinema. In turn, contemporary cinema asks its audience to be astonished at its special effects, and to reflect on the way special effects films have become venues that display developments in new film technology with each new offering attempting to advance and outdo effects that have gone before.

Within the virtual realm of the computer, illusions conjured by the imagination appear to be magically brought to life.[14] Morphing software has combined with other new technologies to conceal the flaws often associated with traditional film effects technology such as blue screen and composites. Since the release of *Willow* in 1988, computer-generated morphing effects have further blurred the line between reality and illusion: each image cross-dissolves fluidly into the next, adding to the illusion of realistic transformations. The more sophisticated the technology becomes, the more photo-realistic the illusions become—and the more the industry itself undergoes a metamorphosis as it moves with greater confidence into the digital realm. For example, laser-based Pixar scanners devised in the mid-1980s allowed morphing to progress smoothly without visible signs of digitization. This cutting-edge technology was aided further in the late 1980s by Photoshop, software that dramatically affected the quality and ease of image manipulation and was used extensively in *The Abyss*.[15] To create the illusion of the transforming pseudopod water creature, stop motion animation was replaced by morphing software. However, the technology and software required to produce the realistic morphing of the water creature not only included the use of new scanners and applications such as Photoshop but also necessitated the expansion of ILM's computer graphics department. At the time, the young Silicon Graphics had been responsible for producing computer workstations for ILM during production of *The Abyss;* by 1993 the company had made profits of more than a billion dollars.[16] Specific technological means were required to create a specific effect (morphing), and in turn, this fantastic transformation within the film's fictional realm led to the further production of the new technology, of new special effects departments and companies, and of a multi-billion-dollar industry. Thus, morphing technology—and computer graphics in general—have been used to make the fantastic seem real, and in the process, film illusions have metamorphosed new technologies and new film industries.[17]

Furthermore, as previously mentioned, the horizontal integration of the entertainment industry has witnessed the migration of blockbuster special effects directors like Spielberg, Lucas, and Cameron into the realm of the participatory theme park attraction—which has itself undergone a radical transformation in the last two decades as a result of these links with both film and computer technology. Lucas, for example, was responsible for *Star Tours* and the *Extra "Terror"restrial Alien Encounter* at Disneyworld, Florida; Spielberg orchestrated the *Jurassic Park* ride at Universal Studios, Los Angeles; and Cameron masterminded the *Terminator 2: 3D* experience at Universal Studios, Florida. Again, fantastic effect paves the way for technological advancement—both within and beyond the film industry. Each new effect marks a turning point that morphs film technology, computer technology, and traditional media forms and takes them to new limits; and each advance is accompanied by redefined audience relationships.

Morphing Magic and a Cinema of Astonishment—"Watch Me Pull a Dinosaur Out of My Mega-gigabyte Hard Disk"

In his article "The Illusion of the Future," Gregory Soloman discusses the "effects Renaissance" of the post-1976 era. He argues that "nowhere is the arrogance of [that] era more manifest" than in the technological advances made by ILM. Films such as *Terminator 2* are viewed as "empty" special effects films: the appearance of the T-1000—and the morphing required to create him—"crassly announces itself as an effect." Soloman sees this spectacle as serving a hollow purpose because it serves no "higher" narrative function.[18] Assuming that spectacle cannot produce "meaning" independently of narrative, Soloman puts forth a view that is dependent on the paradigm of the "classic realist text." As a result, he fails to recognize that this special effects spectacle operates according to a different kind of logic. Indeed, Soloman completely misses the point. Having far more in common with the "attractions" tradition of cinema than with the classic Hollywood tradition, the "crass" announcement of the effect as an effect is precisely the point. Contemporary effects cinema is a cinema that establishes itself as a technological performance, and audiences recognize and revel in the effects technology and its cinematic potential. Rather than centering the action solely around a story, this is a cinema that emphasizes display, exhibitionism, performance, and spectacle. In this regard, it belongs to and continues a tradition primarily concerned with attaining specific kinds of sensory responses from its audience, responses associated with what Tom Gunning has called an "aesthetic of astonishment." In such a cinema, not only is attention diverted away from narrative, but—as in the *Terminator 2: 3D* experience—the films also focus attention back on the artifice of their worlds, evoking in us states of delight that lure us into the attraction through their performance.[19]

Despite embracing technology never seen before, blockbuster effects films of the 1990s share many features with earlier phases in the history of the cinema, in particular, the cinema of the late-nineteenth and early-twentieth century—a cinema Tom Gunning has referred to as the "cinema of attractions."[20] As Gunning briefly states, the special effects renaissance of the "Spielberg-Lucas-Coppola cinema of effects" has much in common with the early cinema of attractions.[21] As with the blockbuster "cinema of effects," the earlier cinema of attractions was not temporally driven; rather, it was a cinema of exhibitionism that delighted in opening up spaces in which the spectator could marvel at the film illusions and their methods of construction.[22] Narrative concerns and characterizations competed with, and were often overpowered by, spectacle and performance. Indirect forms of spectator address were replaced with more active, participatory forms of direct address and engagement. Following this tradition, in the 1990s cinema of attractions, a relationship is established between representation and spectator: one that is intent on leaving the spectator in a state of wonder both at the representation and at the skill and technical mastery that lies behind its construction of the represented spectacle. Special effects spectacles like *Jurassic Park* and the *Terminator 2: 3D* attraction at Florida's Universal Studios construct realistic illusions and spatial relations that invite the audience perceptually to blur the boundaries that separate illusion from reality. In doing so, the amazed audience ponders the technological metamorphosis that the cinema and related crossover entertainment media such as theme park attractions have undergone.

As is the case with earlier forms of attraction entertainment, the special effects forms of the last two decades (especially of the 1990s) have a great deal to do with acclimatizing audiences to different forms of visual engagement that have resulted with the emergence of new digital technologies. In presenting us with performances of special effects spectacles that are intent on showing off digital technology's illusionistic capacity, examples like *Terminator 2: 3D* are emblematic of periods of technological transition and of a society that is itself in the very process of technological metamorphosis. Such special effects displays reflexively engage the audience in their processes of construction.

The self-reflexive attitude in the use of special effects is quite common in contemporary cinema. Indeed, within the space of thirty years, *Star Wars* serves as its own reference point and dramatizes the extent to which such reflexivity goes given the film's rerelease in 1997 in a newly digitized and "enhanced" version. True to the attraction tradition's performative nature, *Star Wars* 1997 is both an homage to the changes that have occurred in computer graphics over the last thirty years since (and due to) *Star Wars* 1977 and a virtuoso performance that takes a bow for the "improvements" in special effects the film has introduced. Just as *Star Wars* 1977 marked a transitional moment in Hollywood cinema, so *Star Wars* 1997 marks a new step toward the cinema of the next century. All the while, in these examples, the films perform for an audience, and the performance centers around special effects technology and its illusionistic potential.

As with early cinema, the spectator is engaged in an ambiguous relationship to this spectacle—a relationship Brooks Landon has called an "aesthetic of ambivalence."[23] Crucial to this relationship is the simultaneous acceptance of the fantastic illusion as *both* a technological achievement *and* a realistic, alternative reality; thus the effects technology is both exposed and disguised. Embodied in this ambivalence is a clash of opposites that connects the rational and technological with the irrational and emotional. The scientific and the rational collide with the fantastic and emotional in a bizarre union explicitly figured by (and as) special effects.

Albert La Valley has characterized special effects as being used to "show us things which are immediately known to be untrue, but show them to us with such conviction that we believe them to be real."[24] In particular, morphing effects dramatically figure and narrativize the function of special effects spectacle in contemporary cinema in that they depend on a transformation that draws us into an illusion. We remain astounded at the effortless magic of the transformations we see before us, yet these very transformations also remind us that they are special effects and ask us to be astounded at the technology that produces such magic. This combination of the irrational with the rational is experienced as the ambivalent tension that tugs at the spectator: the oscillation between emotion and reason—between the sensory and the logical. As Vivian Sobchack states, the "special effects [that] have always been a central feature of the Science Fiction film . . . now carry a particularly new affective charge and value."[25] Meaning, emotional responses, and sensory involvement lie beyond the narrative in the special effects spectacle itself. On the one hand, this affective response almost denies or disguises technology's technological nature; yet on the other, it is the technological effect that makes possible the affect. The concern for evoking states of amazement goes hand in hand with the crafted manipulation of spatial perceptions—and with the active negotiation of the spectator in relation to this environment.

The ambivalence present in film technology's ability to produce "movie magic" is expressed effectively by Martin Scorsese, who states:

> There has always been a magic to the movies. We all know, of course, that movies are the product of science and technology. But an aura of magic has enveloped them right from the beginning. The men who invented movies—Edison, Lumière, and Méliès—were scientists with the spirit of showmen: rather than simply analyze motion, they transferred it into a spectacle. In their own way, they were visionaries who attempted to convert science into a magical form of entertainment.[26]

This association between cinema and magic is not surprising given the cinema's early connections with the magic tradition, and it is a description that has persisted throughout the history of the cinema.[27] In many respects, although today's film technology may be transforming at a dramatic rate and is radically different from that of early cinema, its fundamental concern with constructing magical illusions out of the more rational and scientific realms associated with the technological remains similar. Some of the clearest parallels are to be found between 1990s special effects and precinematic and early cinematic magic performances that owed their illusionistic displays to the magic lantern and other devices. During the nineteenth century, for example, magic performances often served a dual function: they generated an affective response of awe and wonder from the audience so as to stress a more scientific concern with unveiling these fantastic and magical illusions as scientifically constructed and false. As with contemporary special effects cinema, the aim was similar: to lure the audience into embracing the illusion *as reality* in order to expose the affective nature of the illusion *as fabrication*—as the product of technological effect, not magical affect.

Magicians played an important role in the early period of film at the end of the nineteenth and the beginning of the twentieth century. The illusions of magicians found their way onto the screen in numerous ways. In addition to film screenings shown during magic performances, and the production of films depicting magicians' performances, tricks of illusionism so central to the magic performance were given new expression in "trick films."[28] Trick films used the film medium to produce "magic" through means specific to the film medium—for example, through effects such as editing and double exposures. However, magic was also transplanted into film in more direct ways; the transformations, beheadings, disappearances, and ghostly apparitions so popular in trick films owed a great deal to the optical illusions that nineteenth-century magicians achieved previously with magic lanterns. Paralleling current morphing effects in cinema, the effect of metamorphosis, in the nineteenth century, had been a magic lantern optical effect central to the magician's performance. Thus, at the turn of the century, metamorphosis as a cinematic optical effect became an important part of the film magician's repertoire.[29] Many of the metamorphosis trick films exploited the cinema's inherently contradictory nature: its capacity to produce the irrational and the magical in real images through rational and empirical technological means.

Georges Méliès in particular integrated film and magic by expanding the magician's illusionism with the magic lantern into the "trick film" that initiated the special effects tradition. Himself a magician, he purchased the famous Theater Robert-Houdin in 1888 and, soon after, produced a series of "transformation" films including *The Spiritualist Photographer* (1903), which showed a girl in front of a canvas who was then metamorphosed into a painting, then back again.[30] In *The Man with the Rubber Head* (1902) an experimenter (played by Méliès) inflated his own head with a hose; the illusion of the head inflating was achieved through a series of superimposed images. Erik Barnow views the primary concern of this film not as the fantasy of the transforming head but rather as a technical challenge; its achievement asks the audience to be astounded at the technological prowess that makes possible the "morphing" of the head and its realistic appearance.[31] Much like the startling metamorphosis of the Cyberdyne logo into the metal-morphing T-1000 in *Terminator 2: 3D,* such scenes of fantastic transformations were an impressive means of allowing film technology to perform itself. As Gunning suggests, true to the attractions tradition, films addressed and acknowledged the audience by setting themselves up as a technological performance; hence claims to the supernatural were undercut: "Méliès invites technical amazement at a new trick rather than awe at a mystery."[32]

Morphing technology of the contemporary era serves a function similar to that of the nineteenth century: it allows the effect of morphing to take place in an illusionistic display in order to show off the *autoeros* of its own technology. In the process, the seemingly opposite terms "magic" and "technology" often merge. Discussing the morphing effects in *The Abyss,* for example, James Cameron made the following comments:

> Arthur Clarke had a theorem which stated that any sufficiently advanced technology is indistinguishable from magic. And that's how it's supposed to be—for the audience. . . . The audience response to the [pseudopod morphing] sequence was overwhelming. They got the joke; they understood intuitively what was magical about the scene. They were seeing something which was impossible, and yet looked completely photorealistic. It defied their power to explain how it was being done and returned them to a childlike state of entertainment. The sufficiently advanced technology had become magic to them.[33]

In addition to acknowledging the place of *The Abyss* within an attractions tradition in the way the audience is granted active engagement with the special effect ("They got the joke"), Cameron also suggests that an understanding of the "magical" elements of a scene simultaneously implies an understanding (or at least recognition) of the technology that implements it. Spectators are placed in an ambiguous relationship to the screen in that they are invited both to be immersed in and to understand the illusion (the magic) as a reality, and in the methods used to construct that illusion that ruptures its reality. To embrace the technology, the audience must first believe in the fantastic illusion that is conjured. The technology must be both disguised and visible.

Underlying Cameron's comments is a feature central to spectatorship of the contemporary effects film: that is, the spectator's tentative state of uncertainty while in the midst of a game of perception is about "seeing something which [is] impossible, and yet look[s] completely photorealistic." Whether magic transformations, early trick film metamorphoses, or computer-generated morphing—the tricks have always been there, but the technology has changed. The magic lantern morphed into film, and film is now morphing into digital technology. Indeed, the concept of morphing is a useful metaphor that expresses the dynamic interchange that exists not only between spectator and representation in contemporary entertainment media but also between the different media. The special effects illusion—whether it be a morphing cyborg, a computer-generated dinosaur, a 3-D T-Meg, or a digitally processed 70 mm surround image—invites us to embrace its illusionistic universe in real terms.

Perceptions of Realism—Morphing the Reality Experience

Stephen Prince suggests that the fundamental paradox of digitally dependent cinema is its ability to create "credible photographic images of things which cannot be photographed." Computer images thus challenge traditional film theory's grounding on a model of realism that has been tied to "concepts of indexicality" based on film as understood in photographic terms.[34] Computer-generated images force a reevaluation of this tradition because these images have no profilmic referent or source of origin in the real world. In computer-generated images such as the dinosaurs in *Jurassic Park* and the morphed creatures in *The Abyss, Terminator 2,* and *Terminator 2: 3D,* "no profilmic event exists to ground the indexicality of [their] image."[35] Thus contemporary entertainment media open up a different kind of relationship between the spectator and the image. In light of the digital transformation, what Prince actually suggests is that film theory must undergo its own metamorphosis—one that accounts for and responds to digital imagery and addresses the issue of the increased obfuscation of the barriers that separate reality from illusion.[36] More important, he suggests that film theory "has construed realism solely as a matter of reference rather than as a matter of perception as well."[37] Given this stage in the cinema when "unreal images have never before seemed so real,"[38] this position needs to be reevaluated:

> Before we can subject digitally animated cinema and processed images, like the velociraptors stalking the children through the kitchens of

> Jurassic Park, to extended meta-critiques of their discursive or ideological inflections (and these critiques are necessary), we first need to develop a precise understanding of how these images work in securing for the viewer a perceptually valid experience which may even invoke . . . now historically superseded assumptions about indexical referencing as the basis of credibility that photographic images seem to possess.[39]

Film theory fails to account for the paradox that while images may be "perceptually realistic," they are "referentially unreal."[40] Indeed, a great many contemporary blockbuster special effects films reflexively and playfully not only allude to but depend on this paradox. As mentioned previously, since the early film era, technology has aimed at producing perceptually convincing environments. Now this illusion as a perception of reality appears to be more seamless. In a variety of different ways, many contemporary films and related entertainment media perform their games of vision around this notion of constituting illusions as realities. Morphing is an interesting manifestation of this game because in the fluid transformations it produces from one image to the next, it asks that we perceptually accept as real illusions we know cannot be true.

To return to the example of *Terminator 2: 3D,* in addition to employing morphing software and the figural image of the morph, the attraction reflects the way contemporary entertainment media operate according to a morphing metaphor. The attraction lures us into its various levels of reality by displaying (showing off) and exchanging a variety of technological forms and effects—in the process setting itself up as a new kind of techno-spatial-perceptual experience. Participatory techniques are taken to excess: not only are there wider screens, a greater surround sound system, state-of-the-art digital effects, and 3-D technology never witnessed before, but all these technologies combine in an assaultive way that integrates us with both the technology that produces the illusion and the illusion itself. We, in a sense, become morphed into the illusion through the numerous technologies that bombard our senses.

Like the magic shows of the nineteenth century, early cinema, and 1950s wide-screen and 3-D formats, *Terminator 2: 3D* lures us into accepting its illusion as reality. It creates levels of reality that perceptually collapse the frame that separates us from the illusion. This is achieved both through an illusory invasion of *our* space (through 3-D) and through a wide-screen illusion that plunges us into *its* space. In the 1990s, however, computer graphics control every level of production, ensuring the successful, hyperreal articulation of this illusion as reality. Throughout the entire attraction, our sense of reality morphs as we accommodate the various games of perception that are thrust upon us and that we play. When the T-1000 makes his first morphed, 3-D appearance, not only does his screen form as the Cyberdyne logo morph into the chrome man, but the digital metamorphosis also slips into a theatrical metamorphosis: the liquid-metal, computer-generated effect on the screen magically morphs into a live theatrical effect of a live actor onstage and then back again to digital film effect on the screen. We are taken on a journey through different possibilities of "astonishment aesthetics" as the "reality" of the theatrical space of the audience/Cyberdyne complex, the actors, and T-70 cybots performing live around us interweaves with the filmed, videoed, and digitized realities of wide-screen, 3-D, video, and computer images. Moreover, the filmed realities contain further layers that reflect on different constructions of perceived realities. The result is that an interplay occurs between film and digital traditions, one that suggests that the incorporation of the digital into film has "improved" or "advanced" the audience's perception of reality (especially in relation to the perceptions provided by traditional film-viewing experiences). The single twenty-three-by-fifty-foot screen (which suggests more conventional cinematic viewing) transforms to display the morphing effects of the T-1000. For the first time in film history, 3-D technology integrates with computer technology to expand the audience's perception and to convince them that screen space is really collapsing into theater space. The screen expands to triple its length, but the traditional wide-screen experience also introduces another computer-generated spectacle—the T-Meg. This liquid-metal, insectlike creature fills the 180-degree space (which surrounds our vision and therefore creates the illusion that we have entered representational space) and then lunges at the audience in all its morphing, 3-D glory (thus reversing the previous movement by appearing to enter our space).

All these "realities" intermingle: actors from (within) the screen enter into the space of the audience, and the space of the audience appears to become one with the space of the screen (which is what happens when we enter the elevator with the Terminator and John); effects on the screen thrust themselves forcefully into the audience's space through the combination of 3-D and morphing effects, and through theatrical effects such as the sprays of water that hit the crowd when the T-Meg splatters into millions of pieces as it comes straight at us. Computer graphics, 3-D, and wide-screen technologies combine to construct the illusion of a breakdown of the spatial boundaries that separate us and reality from the representation, perceptually collapsing the theatrical frame of the stage. Throughout the entire attraction, however, there still remains an undeniable sense that this convincingly real representational space is also being set up so that the audience may admire it as a multitechnological feat. But unlike the competing 3-D attractions at Walt Disney World *(Honey, I Shrunk the Audience* and *Jim Henson's Muppet Vision 3D), Terminator 2: 3D* doesn't round off its performance with the theatrical closing of the stage curtains. This attraction signals its difference by presenting a performance that is about the removal of the curtain—the removal of the barrier that separates us from the special effects fabrication. Yet that it achieves this so masterfully sets up yet another invisible curtain: one that is drawn to signal closure in our minds—a closure experienced during those moments of deadly silence and stunned amazement that accompany the literally explosive end of the "film." The attraction/film literally ends with a bang as the T-800 blasts the Cyberdyne Complex of 2029 to smithereens. The effects of this explosion are not merely felt perceptually (while remaining contained by the frame of the screen). They are also felt in quite real (yet theatrical) terms through the vibrations we feel under our seats, through the heat that warms our bodies, and through the sea of smoke that veils our vision as it drifts through the auditorium. The silence is soon followed by the audience's tumultuous applause as we acknowledge the technological performance that has just been witnessed. Prince states:

> When the velociraptors hunt the children inside the park's kitchen in the climax of *Jurassic Park,* the film's viewer sees their movements reflected on the gleaming metal surfaces of tables and cookware. These reflections anchor the creatures inside Cartesian space and perceptual reality and provide a bridge between live-action and computer-generated environments.[41]

This perceptual grounding in a Cartesian environment found in *Jurassic Park* (and more interestingly and challengingly in *Terminator 2: 3D*) is, however, also intent on astounding us with the status of such illusions. The game is one that flaunts film's capacity for making a reality out of an illusion—and it leaves us in a state of awe before cinema's ability to present us sensorially with these hyperreal constructions that invite us to embrace them in such real terms while asking us to be amazed at the methods of their construction. Illusion metamorphoses into reality, and reality into illusion. Magic morphs to science, and science to magic. The result? Entertainment forms that are at once magically scientific and scientifically magical.

Notes

1. George Lucas, dubbed by many the "father" of contemporary effects cinema, has anticipated that by the turn of the century, film technology will be replaced entirely by digital production. Lucas was intent on producing all the *Star Wars* prequels using all-digital technology; the technology exists, and the image is of superior quality, but the only problem is that a high-quality 70 mm image would not be ready in time for the prequel production. For more information on the *Star Wars* series, rereleases, and prequels, see the special *Star Wars* issue of *Cinescape* 3, no. 3 (1996), and *American Cinematographer* 78, no. 2 (February 1996).

2. Contemporary blockbuster special effects cinema and attractions owe a great deal to film spectacle of the 1950s. This decade was characterized by its greater emphasis on spectacle, special effects, and more invasive spectator/screen relations, including wide-screen formats such as Cinerama, Cinemascope, and Todd-AO, as well as 3-D cinema. For a detailed account of the changes that occurred in the fifties, see John Belton, *Widescreen Cinema* (New York: Columbia University Press, 1992).

3. For technical information on *Terminator 2: 3D Battle across Time,* see Sydnie Suskind, "Hollywood Bytes," *Cybersurfer* 7 (October 1996): 28–40; and Ron Magid, "Irresistible Force," *Cinescape* 3, no. 3 (1996): 23–29.

4. Similar cutting-edge effects are also present in theater design: new theatrical rigs and motors designed by Scenic Technologies allow props like the 1,500-pound Harley-Davidson to plunge smoothly onto the stage on cue and in perfect synchronicity with events taking place on-screen.

5. Adding to the sensation of the collapse of illusion into reality is the theatrical addition of sprays of water and smoke that integrate us into the action and atmosphere on the screen and in the theater.

6. "*Terminator 2: 3D* Fact Sheet," http://www.usf.com/new/news/press/factt2.html, 1 September 1997.

7. In addition to showcasing its state-of-the-art effects, the attraction that is *T2: 3D* sets up an effects performance that is intent on competing with the other attractions contender in Orlando, Florida: Walt Disney World. As we're ushered into the auditorium, it becomes clear that the Cyberdyne host who introduces us to the company's "cybotics" (as opposed to Disney "animatronics") is a painfully convincing parody of the Disney "have a nice day" personnel. The Disneyfied host's excessive performance reaches a dramatic crescendo when the T-1000 morphs from the screen behind her and (in actor form) approaches her onstage and kills her—just at the point when the audience has had about as much as we can take of her. The result? The crowd applauds furiously in appreciation. The Universal

attraction establishes itself as a very different experience from the attractions provided by the competitor some miles down the road.

8. In his discussion of entertainment cinema of the post-1977 period, Larry Gross, in "Big and Loud," *Sight and Sound* 5, no. 8 (August 1995): 6–10, states that the blockbuster effects phenomenon (which he calls the "Big Loud Action Movie") marks a new form of films that are "blatantly in love with their own technology." Spectacle is focused on at the expense of narrative and concerns with character development.

9. As explained by Tom Gunning, Méliès's trick films were concerned more with the display of magical attractions than with the development of plots and characterization; Méliès's approach to filmmaking owed a great deal to his background as a magician. See Tom Gunning, "Phantom Images and Modern Manifestations: Spirit Photography, Magic Theater, Trick Films, and Photography's Uncanny," in *Fugitive Images: From Photography to Video,* ed. Patrice Petro (Bloomington: Indiana University Press, 1995), 58. Also see Eric Barnow, *The Magician and the Cinema* (London: Oxford University Press, 1981).

10. Numerous writers situate the turning point in the effects rebirth of contemporary cinema around 1977. The success of *Close Encounters* and *Star Wars* revived the industry's (and audience's) fascination with the spectacle of special effects. Although Kubrick's *2001: A Space Odyssey* (1968) had earlier brought to the theater the wide-screen experience and the cutting-edge effects of Douglas Trumbull, arguably, that thoughtful film was not as concerned with aligning the spectator's sensory experience with the effects themselves in an all-embracing, mass-appeal entertainment spectacle. See Vivian Sobchack, *Screening Space: The American Science Fiction Film* (New Brunswick, N.J.: Rutgers University Press, 1997); Joseba Gabilondo, *Cinematic Hyperspace: New Hollywood Cinema and Science Fiction Film: Image Commodification in Late Capitalism* (Ann Arbor: University Microfilms, 1991); Gregory Soloman, "The Illusion of the Future," *Film Comment* 28, no. 2 (March–April, 1992): 32–41; and Gross.

11. Mark Cotta Vaz, *Industrial Light and Magic* (New York: Del Rey, 1996), 9, 91. Exceptions like *2001* aside, such wide-screen and special effects technology was not integral to the entertainment experience in the 1960s and early 1970s.

12. The effects devised through the motion-control camera (called the Dykstraflex after its inventor, John Dykstra) were the first that were the result of a computer controlling the movement of the camera.

13. For more detailed accounts of the advances in digital cinema, and the effects in the films, see Don Shay, "Dancing on the Edge of *The Abyss*," *Cinefex* 39 (August 1989), the issue on *The Abyss;* Jodi Duncan, "A Once and Future War," *Cinefex* 47 (August 1991) (entire issue on *Terminator 2*); James Cameron,

"Effects Scene: Technology and Magic," *Cinefex* 51 (August 1992): 5–7; Suskind; and Cotta Vaz.

14. Computer-generated special effects have none of the cumbersome and practical limitations associated with traditional effects technology. In interviews on his work with digital effects, George Lucas, for example, waxes lyrical about the creative possibilities opened up by this technology, possibilities that allow the filmmaker's imagination free rein. See Kevin Kelly and Paula Parisi, "Beyond *Star Wars,*" *Wired* 5, no. 2 (February 1996): 160–66, 210–12, 216.

15. Cotta Vaz states: "A major breakthrough software (which would be published by Adobe and go on the commercial market in 1990) was Photoshop. Designed to manipulate pictures, Photoshop received its first extensive ILM use for *The Abyss* (1989), helping to manipulate digitized photographic data of background plate sets and creating a digital model in which to integrate a computer-generated water creature. . . . Photoshop edits digitized pictures, allowing computer artists to perform such image-processing functions as rotating, resizing, and color-correcting images and painting out scratches, dust, and other flaws" (117); the result is, of course, that the effect is "disguised." On details of the effects required to produce the water creature, see Cotta Vaz, 115–17, 194–200; Cameron; and Shay. In the wake of 3-D computer graphics employed in *The Abyss,* films that followed took three dimensionality one step further. Although the pseudopod was the first synthetic character to make an appearance in the cinema, it was soon to be upstaged by the first ever digital main character on whom the plot depended: the liquid Terminator T-1000 of *Terminator 2: Judgment Day.* For more on the effects of *Terminator 2,* see Cotta Vaz and Duncan.

16. Cotta Vaz, 195.

17. The initiator in the special effects house boom was Lucas's immensely successful Industrial Light and Magic. The production house was originally formed in 1975 to create the effects for *Star Wars* and then became an independent force that has played a major part in creating some of the most innovative effects witnessed in films of recent years, including *Willow, The Abyss, Terminator 2,* and *Jurassic Park.* In a similar manner, *Terminator 2*'s effects illusions triggered the successful expansion of numerous effects production houses—including director James Cameron's own Digital Domain.

18. Soloman, 32. The spectacle that is the T-1000 does, of course, serve a central narrative function—one pivotal to science fiction thematics: the projected fear of technological advancement as the annihilation of human nature.

19. Although Soloman mourns the T-1000 as a mere "knockoff" of the Silver Surfer comic book character, the T-1000 is also an allusion to the militaristic policemen who populated George Lucas's *THX 1138* (1971). Thus the T-1000 effect

is an homage to the pre–*Star Wars* era of science fiction, but it is also asking the audience to acknowledge the effects advances introduced in the post–*Star Wars* era.

20. Tom Gunning, "The Cinema of Attractions: Early Film, Its Spectator, and the Avant-Garde," in *Early Cinema: Space, Frame, Narrative,* ed. Thomas Elsaesser (London: BFI, 1990), 61. In addition to its pre-1907 manifestation, the attractions tradition has been present throughout the history of the cinema in "performative" and "spectacle" genres such as epics, musicals, comedies, horror, and science fiction. In particular, the aesthetics associated with the cinema of attractions dominate during three phases of the cinema: the pre-1907 period (as outlined by Gunning), the 1950s (and the success of wide-screen and more spatially invasive spectacle), and the blockbuster cinema of the 1980s and 1990s (in its emphasis on both action and special effects).

21. Gunning, "The Cinema of Attractions," 61.

22. Tom Gunning, "'Now You See It, Now You Don't': The Temporality of the Cinema of Attractions," *Velvet Light Trap* 32 (fall, 1993): 6.

23. Brooks Landon, *The Aesthetics of Ambivalence: Rethinking Science Fiction in the Age of Electronic (Re)production* (Westport, Conn.: Greenwood Press, 1992).

24. Albert J. La Valley, "Traditions of Trickery: The Role of Special Effects in the Science Fiction Film," in *Shadows of the Magic Lamp: Fantasy and Science Fiction in Film,* ed. George Slusser and Eric S. Rabkin (Carbondale: Southern Illinois University Press, 1985), 144.

25. Sobchack, 282.

26. Scorsese, quoted in David Robinson, *From Peep Show to Palace: The Birth of American Film* (New York: Columbia University Press, 1996), xi.

27. It may be seen, for example, in the Industrial Light and Magic logo. The logo is the image of a magician coming out of an industrial gear (though in light of the new digital era, this image has now aged).

28. Many nineteenth-century magic shows—including those of the magicians David Devant, Carl Hertz, and George Méliès—included screenings of films such as those first produced by the Lumières. After the premiere of the Lumières' cinematograph, magicians began investing in film cameras and set up their own productions, often including films of themselves performing magic tricks. The magicians who produced such films included David Devant, Leopoldo Fregoli, Alexander Victor (also known as Alexander the Great), Walter R. Booth, and Harry Houdini, who after visiting Méliès's Theater Robert-Houdin, produced a number of films that captured his acts of illusionism. See Barnow, 27–28, 50, 56–81.

29. See Barnow, 16.

30. The theater had previously been established by the famous magician who became Houdini's namesake.

31. Barnow, 97.

32. Gunning, *The Cinema of Attractions,* 63–64. George Lucas is very conscious of the cinema's connections with the magic tradition. He states: "Almost from the moment film was invented there was this idea that you could play tricks, make an audience believe they were seeing things that really weren't there, stretch the imagination. But this was completely lost by the 1960s" (Cota Vaz, 2). This magic and trickery that special effects are capable of producing is what Lucas wanted to bring back with the establishment of ILM.

33. Cameron, 6–7.

34. Stephen Prince, "True Lies: Perceptual Realism, Digital Realism, and Film Theory," *Film Quarterly* 49, no. 3 (spring 1996): 28.

35. Ibid., 29.

36. Ibid., 31.

37. Ibid., 28.

38. Ibid., 34.

39. Ibid., 36.

40. Ibid., 34.

41. Ibid.

Contributors

Roger Warren Beebe is a Ph.D. candidate in the Program in Literature at Duke University. He is a member of the editorial collective of *Polygraph: An International Journal of Culture and Politics* and is coediting an anthology on rock music and culture and, in his spare time, making films and videos.

Scott Bukatman is assistant professor of media studies in the departments of art and comparative literature at Stanford University. His essays have appeared in *Artforum, Architecture New York, October,* and *camera obscura,* and he is the author of *Terminal Identity: The Virtual Subject of Postmodern Science Fiction* and *Blade Runner.*

Victoria Duckett completed her Ph.D. in critical studies in the Department of Film and Television at the University of California, Los Angeles. Her dissertation focuses on Sarah Bernhardt and silent cinema, but her critical and creative work also includes digital media.

Kevin Fisher is completing his Ph.D. in critical studies in the Department of Film and Television at the University of California, Los Angeles. His dissertation is on the cinematic representation of altered states of consciousness in post–World War II American film. He also builds Web sites and creates digital art.

Joseba Gabilondo holds a Ph.D. in comparative literature from the University of California, San Diego, and is assistant professor in the Spanish department at Bryn Mawr College. His book *Consuming Gaze: New Hollywood, Cyberspace, and the Commodification of Otherness* is forthcoming, and he is currently working on a book on contemporary Basque culture.

Marsha Kinder is professor of critical studies at the University of Southern California School of Cinema-Television. She is also director of the Labyrinth Project, a research initiative to expand the language of interactive narrative funded by the Annenberg Center for Communication. She has written many books and essays, and her most recent publications include *Playing with Power in Movies, Television, and Video Games; Blood Cinema: The Reconstruction of National Identity in Spain* (with a companion CD-ROM); and the forthcoming *Kids' Media Culture.* She also coauthored and codirected an alternative CD-ROM titled *Runaways,* a game for teens that uses morphing to explore issues of gender, sexuality, and ethnicity.

Norman M. Klein is a professor at the California Institute of the Arts and a critic-historian and novelist working in urban studies, film studies, the fine arts, new media, and architectural theory. He is author of *The History of Forgetting: Los Angeles and the Erasure of Memory* and *Seven Minutes: The Life and Death of the American Animated Cartoon.* His forthcoming books are *The Vatican to Vegas: The History of Special Effects* and, as coeditor, *Los Angeles in Turmoil.*

Louise Krasniewicz is an anthropologist and award-winning artist and multimedia producer. She is currently the director of the Digital Archaeology Lab in UCLA's Institute of Archaeology, where she translates the research of archaeologists into interactive new media programs. Her anthropological research focuses on issues of narrative, symbolism, and crisis in American communities as well as issues of digital media representation and virtual reality.

Angela Ndalianis teaches film studies in the cinema studies program at the University of Melbourne. Her areas of research include contemporary Hollywood cinema and its intersections with other entertainment media such as theme park attractions, computer games, and comic books. She is currently finishing a book that links the digital age to the baroque period.

Vivian Sobchack is associate dean and professor of film and television studies at the UCLA School of Theater, Film, and Television. She is the first woman to have been elected president of the Society for Cinema Studies. Her work focuses on film and media theory and its intersections with philosophy, perceptual studies, and historiography. Her books include *Screening Space: The American Science Fiction Film; The Address of the Eye: A Phenomenology of Film Experience;* the edited anthology *The Persistence of History: Cinema, Television, and the Modern Event;* and a forthcoming collection of her essays, *Carnal Thoughts: Bodies, Texts, Scenes, and Screens.*

Matthew Solomon is completing his Ph.D. in critical studies in the Department of Film and Television at the University of California, Los Angeles. He is currently writing a dissertation titled "Stage Magic, Spiritualism, and the Silent Cinema," and he has published on radio and cinema sound in *Quarterly Review of Film and Video.*

Mark J. P. Wolf received his Ph.D. in cinema/television at the University of Southern California. He is assistant professor in the communication department at Concordia University, Wisconsin, and has been pursuing research in digital media. He has published in several journals, including *Film Quarterly, Velvet Light Trap,* and *Spectator,* and has two books in progress: *Abstracting Reality: Art, Communication, and Cognition in the Digital Age* and *The Medium of the Video Game.*

Index